Special Relativity

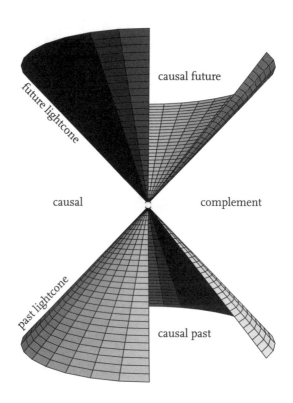

Special Relativity
A First Encounter
100 YEARS SINCE EINSTEIN

Domenico Giulini
Department of Physics, University of Freiburg, Germany

OXFORD
UNIVERSITY PRESS

Great Clarendon Street, Oxford OX2 6DP

Oxford University Press is a department of the University of Oxford.
It furthers the University's objective of excellence in research, scholarship,
and education by publishing worldwide in

Oxford New York

Auckland Bangkok Buenos Aires Cape Town Chennai
Dar es Salaam Delhi Hong Kong Istanbul Karachi Kolkata
Kuala Lumpur Madrid Melbourne Mexico City Mumbai Nairobi
São Paulo Shanghai Taipei Tokyo Toronto

Oxford is a registered trade mark of Oxford University Press
in the UK and in certain other countries

Published in the United States
by Oxford University Press Inc., New York

© Fischer Taschenbuch Verlag in der S. Fischer Verlag GmbH, Frankfurt am Main 2004
English translation © Oxford University Press 2005

Translation and expansion of Spezielle Relativitätstheorie by Domenico Giulini originally
published in German by S. Fischer Verlag GmbH, Frankfurt am Main 2004
Translated by the author
The moral rights of the author have been asserted
Database right Oxford University Press (maker)

First published in English 2005

All rights reserved. No part of this publication may be reproduced,
stored in a retrieval system, or transmitted, in any form or by any means,
without the prior permission in writing of Oxford University Press,
or as expressly permitted by law, or under terms agreed with the appropriate
reprographics rights organization. Enquiries concerning reproduction
outside the scope of the above should be sent to the Rights Department,
Oxford University Press, at the address above

You must not circulate this book in any other binding or cover
and you must impose this same condition on any acquirer

A Catalogue record for this title is available from the British Library

Library of Congress Cataloging-in-Publication Data

Giulini, D. (Domenico), 1959–
Special relativity : a first encounter, 100 years since Einstein / Domenico Giulini.
p. cm.
Includes bibliographical references and index.
ISBN 0-19-856746-4 (alk. paper) — ISBN 0-19-856747-2 (alk. paper)   1. Special
relativity (Physics)   I. Title.
QC173.65.G58 2005
530.11—dc22

2004024330

ISBN 0 19 856746 4 (Hbk)

10 9 8 7 6 5 4 3 2 1

Typeset by Newgen Imaging Systems (P) Ltd., Chennai, India
Printed in Great Britain
on acid-free paper by Biddles, King's Lynn.

Preface

The theory of Special Relativity was undoubtedly one of the most outstanding achievements of 20th century physics, and the year 2005 marks the 100th anniversary of Ein-stein's original publication. In this book I try to explain Special Relativity to an audience of non-expert readers. There are, in my opinion, very few areas in fundamental physics where insights of tremendous depth can be gained by comparatively simple methods, Special Relativity being one of them.

More specifically, I address the following four questions: How did Special Relativity originate? What are its central statements? What are its most significant applications? What is its present experimental status?

Any author speaking to a wider audience on a subject of this nature must be careful to keep the mathematical formalism to a minimum; formulae are often therefore replaced by text or diagrams. On the reader's side this will require, above all, a degree of patience and the desire to understand. I have made no attempts to replace Special Relativity by an easier caricature of itself. Rather, I have followed the dictum often attributed to Einstein: 'Make things as simple as possible, but no simpler'.

To be more precise, the required level of mathematical proficiency is that of a well educated 16 year old. I use elementary algebra, geometry, and trigonometry; the exponential function appearing just once. Elementary calculus (i.e. integration and differentiation) is not required to understand any of the displayed formulae, but in some rare cases it has been more or less implicitly used in their derivation.

The objectives of my presentation are fourfold: To be faithful regarding the foundations; to be up to date, especially regarding the experimental status; to be compact; and, as previously mentioned, to be accessible to the non-expert. Even though there is a vast literature on the subject of Special Relativity, I believe only few texts exist which attempt to meet these challenges simultaneously. It is clearly a matter of compromise. Whether I found the right balance is for the reader to decide.

Last, but not least, I wish to thank Oxford University Press for their excellent cooperation and, in particular, Sönke Adlung for his enthusiasm for this project.

Freiburg, November 2004 *Domenico Giulini*

Contents

1 Origin and significance of Special Relativity 1

2 Historical developments 5
   2.1   The dualistic concept of matter in the 19th century 5
   2.2   The principle of relativity in mechanics 11
   2.3   Is the relativity principle valid in electrodynamics? 20
   2.4   Experiments, contradictions, and consequences 21
       2.4.1 Aberration 22
       2.4.2 Fizeau's experiment 25
       2.4.3 The Michelson–Morley experiment 28
       2.4.4 The FitzGerald–Lorentz deformation
             hypothesis 35

3 Foundations of Special Relativity 38
   3.1   The notion of simultaneity 41
   3.2   Lorentz transformations 46
   3.3   Time dilation and length contraction 51
       3.3.1 Time dilation 51
       3.3.2 Length contraction 54
   3.4   Addition of velocities 58
   3.5   Causality relations 60
   3.6   Aberration and Doppler effect 63
       3.6.1 Aberration 63
       3.6.2 Doppler effect 65
   3.7   Length contraction and visual appearance 67
   3.8   Mass, momentum, and kinetic energy 69

3.9	Probably the most famous formula in all of physics ...	75
3.10	Electrodynamics: Invariance of Maxwell's equations...	80

4 Further consequences and applications of Special Relativity 86

4.1	Atomic physics	86
4.2	Nuclear physics	88
4.3	Elementary particle physics	92
4.4	Daily physics: navigational systems	98
4.5	Science fiction: travel to distant stars?	102
4.6	Outlook on General Relativity	105

5 Closer encounters with special topics 110

5.1	Ole Rømer's measurement of the velocity of light.....	110
5.2	The independence of the velocity of light from the state of motion of the source............	114
5.3	Do superluminal velocities exist?............	117
5.4	The Kennedy–Thorndike experiment............	122
5.5	The Ives–Stilwell experiment	127
5.6	The current experimental status of Special Relativity	130
5.7	Synchronization by slow clock-transport	137
5.8	Aberration and conformal transformations	139
5.9	Transformation formulae for momentum, energy, and force	142
5.10	Minkowski space and the Lorentz group	144

Bibliography 154

Glossary 158

Symbols, units, constants 161

Picture Credits 162

Index 163

·1·
Origin and significance of Special Relativity

The year 1905 is commonly known as Einstein's miraculous year. In that year, the just 26 year old patent clerk Albert Einstein (1879–1955)—at this time still a scientific nobody—published five seminal papers in the prestigious German physics journal *Annalen der Physik*, each of which had a major impact on the future development of physics. (A book containing all five papers in English translation appeared a few years ago [1], now also available as an e-Book.)

In the first paper he proposed his light-quantum hypothesis and used it to explain the photoelectric effect, which brought him the Nobel prize for the year 1921 (received in 1922). Out of the five papers from that year, this is the only one which Einstein himself explicitly ranked as 'very revolutionary'. The second paper was his PhD thesis, in which he derived an analytical relation between the true size of the molecules of a dissolved substance and the viscosity of the solution. Due to its various applications, in particular in petrochemistry, this paper led the citation list of all Einstein papers, at least until the 1980s. The third paper deals with the statistical theory of heat, which Einstein had independently developed and which he used here to explain the phenomenologically well established, but theoretically poorly understood, Brownian motion as statistical fluctuation phenomenon. (By 'Brownian motion' one understands the irregular jittering motion of microscopically small particles in liquid suspensions.) This lent decisive observational evidence to the statistical theory of heat, which at that time was still fairly controversial due to its fundamentally atomistic approach. The fourth paper carries the title 'On the Electrodynamics of Moving Bodies' and contains essentially what we now call *Special Relativity*, henceforth simply abbreviated to 'SR'. Finally, the fifth paper is an addendum to the fourth and contains on less than three printed pages the derivation of probably the most famous formula of physics: $E = mc^2$.

SR is a theoretical framework and not so much a theory of a well defined domain of phenomena, though it owes its existence to such a theory, as is already apparent from the original title given above. The electrodynamics of moving bodies was one of the big issues in experimental as well as theoretical physics of the late 19th and early 20th century, which found itself increasingly entangled in difficulties, up to plain inconsistencies, until Einstein cut the Gordian Knot in a surprising fashion. Despite the indubitable ingenuity of Einstein's solution, it would be quite inappropriate to assign all credits for this development to him alone. Retrospectively, SR seems palpably close in 1905, after all the preliminary works of Voigt, Hertz, FitzGerald, Lorentz, Larmor, and Poincaré. But apparently it needed an unprejudiced newcomer to take the final step. This step did not consist in a still more refined improvement on that part of the theory which connects to the phenomena, but rather in a fundamental scrutinization of apparently well established notions concerning space and time, like 'distance', 'duration', and 'simultaneity'.

Since *all* physical processes take place in space and time, the revision of space-time concepts initiated by Einstein eventually affects all of physics. Hence, even though SR owes its existence to specific issues in electrodynamics, it is not logically tied to it. Except for gravity, which is described by General Relativity, all fundamental interactions—electromagnetism, the strong or nuclear interaction, and the weak interaction—are nowadays described by theories which obey the axioms of SR. In particular this is true for the so called 'Standard Model' of elementary particles, in which all interactions but gravity are mathematically combined. Without SR modern high-energy particle physics would be unthinkable.

But not only in particle physics, which is somewhat remote from everyday experience, is SR of fundamental importance. For

Origin and significance of Special Relativity

example, the modern technologies of geodesy and navigational systems are essentially based upon the principles of SR, in particular upon the *universality of the velocity of light*. By this one means that the velocity of light measured by an observer is independent of the state of motion of either the source or the observer. A modern example is given by the satellite-based navigational system GPS (Global Positioning System), which allows the determination of one's position from the travel-times of electromagnetic signals sent out by the satellites. The universality of the velocity of light is crucial in order to unambiguously convert these times into lengths and hence positions. It can hardly be overstressed in what a fundamental fashion this universality contradicts the firmest views on space-time measurements taken upto and into the 20th century: How can a propagation process have unchanging velocity, even if one approaches or flees it at arbitrary speeds?

For us, whose intuitive understanding of space-time relations have been developed and trained in everyday life, certain results of SR have undeniably certain paradoxical aspects to it. Here we wish to strictly distinguish between 'paradoxical', meaning 'being against a held opinion', and 'contradictory' in the logical sense: SR does not contain any logical inconsistencies; it just doesn't concur with all our expectations, which are based on extrapolations. Physically this situation arises due to the almost fantastic magnitude of the velocity of light, whose exact value in units of kilometre (km) per second (s) is given by

$$c = 299\,792.458 \text{ km/s}. \tag{1.1}$$

(The exactness being simply due to the fact that since 1983 the 'kilometre' is *defined* in terms of the 'second' and the stated value for c.) This enormous velocity is far bigger than all velocities of material bodies we encounter in daily life. Indeed, up to the astronomical measurements of the second half of the 17th century, it was even undecided whether light might not propagate instantaneously, i.e. with infinite speed. And it required the much refined experimental technology of the 19th century to allow measurements of the speed of light over terrestrial distances. It is therefore absolutely sufficient for our daily life to approximate this velocity by infinity. The click of the switch of a torch and the visual impression of its entire light cone are, to a very good approximation, simultaneous events. But, as we will see, the actual finiteness of the speed of light and its rôle as an upper limit for *all* signal velocities enforces a deep revision of our intuitive notions concerning space-time relations. In particular, the notion of simultaneity of spatially separated events needs to be revised in order for it to make any operational sense in situations where the velocities involved are no longer negligible compared to the velocity of light. At this point it is important to keep in mind that this revision was born out of a real crisis in the foundations of physics, which emerged from experimental facts on one hand and the theoretical notions and their relations on the other. Up to now the space-time concepts of SR have proved extremely useful in understanding physical processes in the absence of gravitational effects. In this realm SR has passed the modern precision test with much bravura and shows no signs so far of any deviations.

·2·
Historical developments

2.1 The dualistic concept of matter in the 19th century

In 1687 Isaac Newton's (1643–1727) *Philosophiae Naturalis Principia Mathematica*, nowadays simply called 'The *Principia*', appeared in print in London. In this monumental work, which influenced the physical discipline of mechanics like no other ever since, Newton laid down a physical theory in mathematical terms which allowed him to describe the motion of heavenly bodies within the very same formalism as terrestrial motions. Quite generally, Newton speaks of 'bodies', which eventually one has to think of as being built from infinitely small, infinitely tough, and never changing parts, which themselves are taken to require no further explanation. With this concept of 'point masses', as they are now called, Newtonian mechanics is able to reduce the motion of complex configurations of such point masses to their simple laws of motion, taking into account the forces between them. A simple though somewhat idealized hypothesis for such mutual forces leads to the notion of the 'rigid body', in which each point mass is held in constant position relative to the others, even if external forces are applied. The spatial configuration of such an ideal rigid body is fully characterized by just six numbers, three for the position of some preferred point of it, like e.g. the centre of mass, and three for the rotational freedom about this point. Clearly such a concept is really to be thought of as an approximation, which is valid as long as the external forces (e.g. gravitational forces) which might act on the extended body are of a much smaller strength than the binding forces which keep the point masses in place (e.g. electrostatic forces). Otherwise the body will deform and eventually disintegrate. This reductionist programme proves extremely useful and leads to an immense wealth of applications, covering the dynamics of a bicycle and planetary motions alike. So if one asks to what material entities Newtonian mechanics is in principle applicable, the answer will be that everything qualifies that can be thought of as being composed of such elementary point masses. If one adopts the viewpoint of naive atomism, one might even conjecture *all* physical phenomena to be eventually reducible to the laws of mechanics.

Really all? Newton also investigated optical phenomena and ventured some hypotheses on the nature of light and its laws of propagation and interaction with matter (in his '*Opticks*' from 1704). But he was not able to formulate a consistent system of concepts and laws comparable to his *Principia*. In fact, Newton assumed light to also consist of particles which could be acted upon by forces, like the gravitational force (which Newton thought could explain the phenomenon of refraction). But this particle theory of light was overturned in the 19th century in favour of the competing notion of light as a wave. This was essentially due to the wonderful experiments of Thomas Young (1773–1823), which proved unambiguously that light interferes, that is, that superpositions of light beams not only result in enhancements of intensities, but sometimes also in attenuations or even total cancellations. This phenomenon cannot be explained within a particle theory, in which superpositions of particle beams will clearly always result in enhancements of local particle densities.

But if light is a wave, that is, a propagating oscillation process, one must ask what it is that is oscillating there. In analogy to water waves on the surface of a lake, in which the water molecules oscillate vertically in space, light should correspond to the oscillation of some hypothetical medium which was termed the 'ether'. This 'ether' should then be able to penetrate all materials in which light can propagate, like e.g. water or glass which, after all, have quite a significant density. Moreover, it has been known since the measurements of the Danish astronomer Ole Rømer (1644–1710),

The dualistic concept of matter in the 19th century

performed in the years 1672–76, that the velocity of light is given by an enormous value, which Rømer first quoted as 220 000 kilometres per second, which is 3/4 of the exact value (1.1) that we know today and which lies pretty close to 300 000 kilometres per second. This extreme value already makes it clear that the analogy between light waves on one hand and waves of deformations of an elastic material on the other will meet substantial difficulties. It is known that the speed of elastic waves grows proportional to the square-root with the strength of the material. As a consequence, the ether's mechanical strength would come out quite fantastically, far beyond that of any known material. On the other hand, as already mentioned, the ether was at the same time required to easily penetrate other material and cause no hindrance to planetary motions. Obviously these requirements do not seem to fit together very well.

In spite of these incompatible properties people maintained the idea of an ether of some sort, albeit without any underlying physical understanding of it. In the absence of any consistent theory of the ether, not only did the wave theory of light seem to be without physical basis, but also it seemed incomprehensible how forces could act over large spatial distances without the assumption of some medium that could physically mediate actions of force. Newton, too, was convinced that such a medium must exist, despite the fact that he made no such suggestions in his *Principia*, where he merely described the precise actions (at a distance) of gravitational forces without contemplating any mechanisms of their transport. But Newton was more frank in his letters. A wonderful passage in one of this famous letters to Robert Bentley, dated February 25 1692/3, reads as follows [2]:

Tis unconceivable that inanimate brute matter should (without ye mediation of something else wch is not material) operate upon & affect other matter wthout mutual contact; as it must if gravitation in the sense of Epicurus be essential & inherent in it. And this is one reason why I desired you would not ascribe innate gravity to me. That gravity should be innate inherent & essential to matter so yt one body may act upon another at a distance through a vacuum wthout the mediation of any thing else by & through wch their action of force may be conveyed from one to another is to me so great an absurdity that I believe no man who has in philosophical matters any competent faculty of thinking can ever fall into it.

Clearly, the same could be said of electric forces, which were much studied at the end of the 18th century by the French physicist Charles Augustin de Coulomb (1736–1806), who proposed a force law for electric point charges—nowadays known as Coulomb's law—in full analogy to Newton's gravitational law.

In 1873 a comprehensive and powerful theory for all electromagnetic phenomena was published by the Scotsman James Clerk Maxwell (1831–1879), who was guided by the more intuitive ideas of the English chemist and experimental physicist Michael Faraday (1791–1867). In order to describe spatial distributions of actions of force Faraday developed the idea of 'field lines', which he separately associated to electric and magnetic actions. At the beginning this might have just been a trick to visualize the spatial distribution of electric and magnetic forces that would be exerted if electric test charges and test currents were placed at various positions in space. But Faraday took a further logical step in taking these field lines more seriously. He endowed them with physical reality, independent of the presence of any test charges and current. Thereby he introduced a new reality concept into physics: the electric and magnetic field. To each point in space one associates an electric and magnetic vector, that is, a direction and a strength. If one accepts this idea, the question is how these fields distribute in space and change in time, depending on the external charges and currents. In particular, it now makes sense to ask for the electric and magnetic fields at locations without charges and currents, i.e. in 'vacuum'. These questions are fully answered by Maxwell's mathematical formalism, whose physical meaning is based, as we wish to stress again, on Faraday's concept of fields. The mathematical theory itself is of great structural beauty. In particular, it implies that in case of time dependent field configurations electric and magnetic fields mutually depend on each other. This dependence is

The dualistic concept of matter in the 19th century

so intimate that it is more appropriate to simply speak of a unified *electromagnetic* field, which now has six components (three electric, three magnetic) per space point at any given time. We will later see that SR makes it manifestly impossible to perform an absolute split between electric and magnetic components.

One of the most impressive achievements of Maxwell's theory was the prediction of electromagnetic waves which could propagate in vacuum. The speed of propagation is also predicted by the theory and turns out to be equal to the speed of light. This suggested that light might be nothing else but an electromagnetic wave and that the laws of optics, like the laws of refraction and reflection, are deducible within Maxwell's theory, a hope that was brilliantly realized during the late 19th century. These achievements, plus the sensational experiments undertaken in 1888 by Heinrich Hertz, who produced and verified the existence and propagation of electromagnetic waves in the laboratory, firmly established Maxwell's theory. With it the notion of a field as fundamental local physical entity was accepted, though it was not yet regarded as a fully emancipated form of matter.

Even Maxwell did not free himself from the idea of an ether. How should one understand the notion of a field at some point at which there are neither charges and currents nor any other form of matter? What does it mean to assign a 'field vector' to a 'point' without any material identity? How can it then be 'attached'? Or put differently: If the field is a quantity of state, *whose* state are we then talking about?

In fact, Maxwell tried to think of the ether in terms of a mechanical model; cf. **Fig. 2.1**. His own theory of the electromagnetic field would then merely be some sort of coarse grained description, valid for phenomena on scales much larger than the typical scales of structures in the ether, like the 'ether vortices'. In that respect Maxwell's theory would then be somewhat analogous to the phenomenological theory of gases, developed also by Maxwell and after him Ludwig Boltzmann (1844–1906) and Josiah Willard Gibbs (1839–1903). They succeeded to reduce typical notions, like 'temperature' or 'pressure', to mechanical notions by assuming a

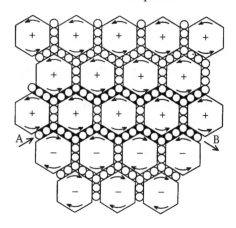

Fig. 2.1 Adaptation of a sketch by Maxwell concerning his mechanical interpretation of an ether.

gas to be nothing else than a large number of tiny and fast moving molecules. **Figure 2.1** is an adaptation of Maxwell's own drawing, showing his mechanical model of the ether. Magnetic fields are produced by 'molecular vortices', which are mutually held in position through charged particles, just like the ring in a ball-bearing is held in place by the balls. Different rotation speeds of neighbouring vortices then result in transport of charges. Other scientists after Maxwell had similar ideas upto and into the 20th century. Even Hertz, who once said that 'Maxwell's theory are Maxwell's equations', devoted the final years of his all too short life to an attempt to give a new and modern analytic formulation of mechanics resting on an axiomatic basis. His hope was that by eliminating dispensable and unclear notions (like, as he thought, that of a 'force'!) the whole setup could be made more rigorous and hence more controllable, and that this would help to eventually describe the ether in a modernized mechanical formalism. But except for an admirably clearly written book, published only posthumously, this and all other programmes to reduce the ether to the laws of mechanics never succeeded.

Consequently a dual notion of matter prevailed upto and into the 20th century. On one hand it comprised the localized 'bodies' of all sorts, which also carried the attribute of inertia and gravity, and which were therefore called 'ponderable' (i.e. 'weighable'). On the other hand there was the ether, spread throughout space including the interiors of bodies, which was the carrier of electromagnetic fields and, in particular, light. The gravitational field, too, was thought to be anchored within the ether, though by that time there was no fully developed theory of the gravitational field comparable to Maxwell's formalism.

2.2 The principle of relativity in mechanics

A central notion in Newtonian mechanics is that of a *force* (despite Hertz). Quite generally one many say that on one hand Newtonian mechanics is about deducing types of motion from known actions of forces, and on the other about deducing laws of forces from observed types of motions. This works through Newton's equation *force = mass × acceleration*, which formally reads:

$$\vec{F} = m\vec{a}. \tag{2.1}$$

Here the arrow over F and a indicate the vectorial, i.e. directed, nature of the quantity. A force and an acceleration have not only a strength, also called the 'norm' or 'magnitude' of the vector, but also a direction. Equation (2.1) then says that the acceleration is parallel to the force and that its strength equals $1/m$ times that of the force. Relative to a *reference system* one can fully characterize a vector by three numbers together with the physical unit. These are called the *components* of the vectorial quantity. Changing the reference system results in a change of components.

According to this, forces are the causes of accelerations, that is, changes of velocities. Since both quantities are vectors, this holds not only with respect to the magnitude but also with respect to the directions. For example, swinging the hammer in a circular orbit at a constant rate still continuously changes the direction of its velocity, though not the magnitude. As a result the acceleration momentarily points perpendicular to the velocity and this is why the athlete needs to exert a strong pull.

The Newtonian formula (2.1) now implies that in the absence of forces accelerations must vanish, that is, velocities must be constant. Since the velocity is a vector, denoted by $\vec{v}$, this means that its magnitude as well as its direction remains constant if no external forces are acting. This is just the statement of the law of inertia:

Law of inertia *A force-free body remains at rest or in a state of rectilinear and uniform motion.*

We add that according to Newton's formula *any* rectilinear and uniform motion is compatible with the absence of forces. The magnitude and direction of the velocity can be freely chosen. Also the position that the body takes at fixed time, say $t = 0$, is totally undetermined by the law. In particular, we have

Mechanical principle of relativity *Two identical physically closed systems whose relative motion is rectilinear and uniform are indistinguishable with respect to mechanical observables of the individual systems.*

The insight into the validity of this principle predates Newton's *Principia* of 1686. It was beautifully pictured by Galileo Galilei (1564–1642) in his 'Dialogue Concerning the Two Chief World Systems' of 1632. There, on the second day, Galileo's alter ego, the Florentine patrician Filipo Salviati, explained it as follows:

Shut yourself up with some friend in the main cabin below decks on some large ship, and have with you there some flies, butterflies, and other small flying animals. Have a large bowl of water with some fish in it; hang up a bottle that empties drop by drop into a narrow-mouthed vessel beneath it. With the ship standing still, observe carefully how the little animals fly with equal speed to all sides of the cabin. The fish swim indifferently in all directions; the drops fall into the vessel beneath; and, in throwing something to your friend, you need throw it no more strongly in one direction than another, the distances being equal; jumping with your feet together, you pass equal spaces in every direction. When you have observed all these things carefully (though there is no doubt that when the ship is standing still everything must happen this way), have the

The principle of relativity in mechanics

ship proceed with any speed you like, so long as the motion is uniform and not fluctuating this way or that. You will discover not the least change in all the effects named, nor could you tell from any of them whether the ship was moving or standing still.

Note that in this example it is essential that *all* physical components of the system are taken alike by the translational motion, in particular also the air that is enclosed by the vessel. Precisely for this reason the experiments are to be performed 'below decks'. If the portholes were opened so that the air could stream through, the butterflies would of course no longer show an isotropic velocity distribution relative to the ship but rather follow preferably the direction of the airflow. We will come back to this picture when discussing the relativity principle in electromagnetism.

At this point we wish to comment on a fundamental aspect in connection with the conventional phrasing of the law of inertia as given above, which is not often expressed sufficiently explicitly. This formulation does not contain any information with respect to which reference system the inertial (i.e. forceless) motion is to be rectilinear and with respect to what timescale it is to be uniform. For example, inertial motion is certainly not along straight lines with respect to a reference system that is rigidly attached to the body of the Earth. In the same fashion, it is in no way uniform with respect to a clock whose rate changes in time relative to a 'normal' clock. Such a clock would not necessarily be useless, as long as its rate is well defined and reproducible. For example, up to the 15th century it was quite customary to divide the day, i.e. the time between dawn and sunset, and the night, i.e. the time between sunset and dawn, each into 12 mutually equal day and night hours, respectively. During summer time day hours were then longer than night hours and vice versa during winter time. Moreover this difference depended on the geographic latitude. There still exist old mechanical clocks at various places along the coast of the Baltic sea with two faces painted on top of each other, one being divided into the day-night hours just mentioned, which are called 'temporal hours', and the other into the 'equinoctial hours', as they were then called, which are the ones familiar to us today. To come back to the law of inertia we may ask: how do we know that the uniformity expressed there is meant with respect to the equinoctial scale and not with respect to the temporal scale? The succinct answer to this is that otherwise the law of inertia would not be true! The point being that the above formulation of the law of inertia is strictly speaking incomplete, and should be completed as follows:

Law of inertia (completed) *A force-free body remains at rest or in a state of rectilinear and uniform motion, if the spatial reference system and the time scale is chosen appropriately.*

In other words, the law of inertia asserts the *existence* of preferred reference systems and time scales with respect to which inertial motion is rectilinear and uniform. Following a suggestion of Ludwig Lange's (1863–1936) [3] such reference systems and time scales are called *inertial systems* and *inertial time scales*. These are by no means uniquely determined by the law of inertia. A time scale remains inertial if to each of its values a fixed value is added or if each value is multiplied by the same (non-zero) number. A reference system remains inertial if it is shifted by a fixed amount, rotated by a fixed amount, or put into rectilinear motion with constant speed. The last operation is called a velocity transformation or simply a 'boost', whose analytic expression we wish to state explicitly. Let the spatial reference system be analytically represented by an orthogonal coordinate system K whose axes we denote, as usual, by x, y, and z. Let t be the time measured in this reference system. Let K' be a second coordinate system whose axes x', y', and z' are pairwise parallel to the corresponding axes of K, and which moves with velocity v along the x-axis of K such that at $t = 0$ the two systems coincide. The inertial time t' measured in K' is taken identical to t. Then we have (see **Fig. 2.2**)

$$x' = x - vt, \quad y' = y, \quad z' = z, \quad t' = t. \tag{2.2}$$

This means that an event whose space-time coordinates with respect to K are (x, y, z, t) has the space-time coordinates $(x - vt, y, z, t)$ with respect to K'. Generally, such formulae which

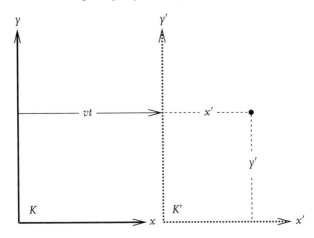

Fig. 2.2 Relation between the coordinates (x, y) and (x', y') of a point • at time t, with respect to the coordinate systems K and K', respectively. K' is moving at constant speed v relative to K in the x-direction. The third dimension (z-coordinate) is suppressed.

relate coordinates and time scales of different inertial systems are called *transformations*. The transformation (2.2) is called a *Galilei transformation*. We wish to deduce an apparently self-evident consequence of it, namely the law for the addition of velocities. To do this we imagine a projectile, which relative to K' moves with the constant velocity $\vec{u}' = (u'_x, u'_y, u'_z)$. This is analytically described by

$$x' = u'_x t', \quad y' = u'_y t', \quad z' = u'_z t'. \tag{2.3}$$

Inserting these expressions for x', y', and z' into (2.2) and solving for x, y, and z yields

$$x = (u'_x + v)t, \quad y = u'_y t, \quad z = u'_z t, \tag{2.4}$$

from which we read off the projectile's velocity $\vec{u}$ relative to K:

$$u_x = u'_x + v, \quad u_y = u'_y, \quad u_z = u'_z. \tag{2.5}$$

This is the classical and intuitively seemingly obvious law of addition of velocities. It merely states that velocities are to be added vectorially, i.e. componentwise. But here it is important to add the remark that this plausible rule is by no means forced upon us by mere logic. The operation of velocity addition is defined physically and need not necessarily be represented by the simple mathematical operation of vector addition. That this is the case in the present context is a non-trivial statement about Newtonian mechanics. We will see that in SR this simple rule gets replaced by something much more complicated.

The temporal development of simple processes can be conveniently depicted in so-called space-time diagrams. These contain a time axis in addition to the spatial coordinates, which is usually depicted in a vertical upward direction. The unique fixing of a point in space-time, also called an 'event', then needs three space and one time coordinate, i.e. four number-valued data. In that sense one says that space-time is four dimensional. The motion in time of a point through space is represented by a directed line in space-time, which one calls the particle's *world line*. The world lines of force-free particles will be straight if and only if the coordinates refer to an inertial reference system and an inertial time scale, in which case the whole four-dimensional spatio-temporal reference system is likewise simply termed 'inertial'. The slopes (i.e. the tangent function of the angle) of these straight lines against the vertical then equal the velocities in that reference system. **Figure 2.3** shows the world lines of two scattering particles A and B. Before the collision event particle A moves along the x-axis in a positive direction and particle B with equal speed in the opposite direction. At time $t = t_Z$ and location $x = x_Z$ an elastic collision takes place after which the particles separate with the same speeds in opposite directions. Since the particles exchange no forces except at the collision point--in particular, they are assumed uncharged—their world lines are straight except for the point of interaction where they suffer a kink. Such elementary and well localized interactions serve for physicists as an operational approximation to what mathematicians call a 'point' in space-time or simply an 'event'. To be sure, no real physical event really defines a mathematical point,

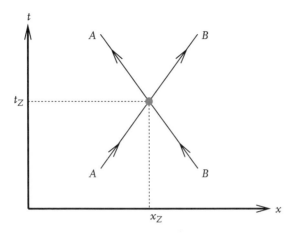

Fig. 2.3 Scattering of two particles A and B in a space-time diagram. The collision event has space-time coordinates x_Z and t_Z.

but rather a small region of finite extent, as depicted in **Fig. 2.3** by the shaded region where the particles meet. But it is nevertheless useful and, under defined conditions, also admissible to think of points as being also physically determined, even if this can at best only be true in an approximate sense. As a side remark we add that even though there is no complete theory of quantum gravity as of today, a heuristic combination of the fundamental principles of quantum theory with general relativity strongly suggest that there exist *fundamental* lower bounds for localizability in space and time. These are given by the so-called Planck length, ℓ_P, and Planck time, t_P, which are expressible in terms of other, more familiar fundamental constants, the velocity of light c, Planck's constant $\hbar$, and Newton's constant G:

$$\ell_P = \sqrt{\frac{\hbar G}{c^3}} = 1.62 \cdot 10^{-35} \text{m}, \tag{2.6}$$

$$t_P = \sqrt{\frac{\hbar G}{c^5}} = 5.40 \cdot 10^{-44} \text{s}. \tag{2.7}$$

Such fantastically short lengths and times are by many orders of magnitude way out of the reach of present achievable resolutions in particle physics. One therefore takes the point of view that, as long as one keeps well away from these scales, it is an admissible idealization to identify fundamental physical events with points. This clearly proves extremely convenient for the mathematical description.

Equation (2.2), which gives the algebraic expression for the Galilei transformations, can now easily be interpreted geometrically, as depicted in **Fig. 2.4**. To understand this, recall that the t-axis is defined to be the set of all events whose space coordinate x equals zero, i.e. the world line of the space point $x = 0$. All lines parallel to it are the world lines of other fixed space points. Likewise, the x-axis is the set of events at time $t = 0$ (i.e. mutually simultaneous) and all lines parallel to it are the mutually simultaneous events for other values of time. Therefore, the t'-axis of the system K' which moves relative to K is just the world line of $x' = 0$. It moves with velocity v in the positive x direction, which means that

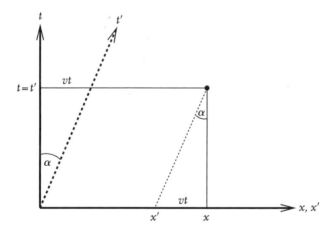

Fig. 2.4 Geometric interpretation of a Galilean transformation. The event ● has coordinates (x, t) with respect to K and (x', t') with respect to K'.

the t'-axis is inclined with respect to the t-axis by some angle α. On the other hand, the x'-axis is identical to the x-axis, for points which are simultaneous in K are also simultaneous in K' and vice versa. This is just expressed by the last equation of (2.2). This seemingly obvious assumption is an expression of an *absolute* (meaning independent of the inertial system) notion of *simultaneity*. We stress that this rests on an assumption and not on any logical necessity. We will see that SR replaces this absolute by a relative notion of simultaneity, which depends on the state of motion of the inertial system. This will result in modified transformation formulae for which, in contrast to **Fig. 2.4**, the x'-axis no longer coincides with the x-axis.

In order to prevent possible misinterpretations at this point, we wish to stress that in principle we are totally free to use *any* spatial reference systems and time scales. There is nothing 'unphysical' or even forbidden about non-inertial systems. Inertial systems are merely *preferred* by the laws of nature, since with respect to them the laws acquire a particular simple form. Only with respect to inertial systems and time scales is Newton's law given in the simple form (2.1). In non-inertial systems there would be additional terms in it, which take account of the 'inertial forces', like the Coriolis force or the centrifugal force. To master concrete situations this flexibility in one's choice of reference systems proves extremely convenient. For example, for terrestrial situations it is convenient to use spatial coordinates which are rigidly connected to the Earth's body and a 'clock' whose hand is the Earth's rotation angle against the Sun, measured from some fixed longitude (e.g. Greenwich). This system is not inertial for various reasons: the spatial system certainly not because of the Earth's intrinsic rotation. Also, the time scale is not inertial since the rotation speed of the Earth against the Sun is not uniform, mainly due to the annual variation in the separation of the Earth to the Sun, but also due to other intrinsic effects, like tidal friction, which let the Earth's rotation speed vary even with respect to the most distant astrophysical objects, like quasars. It needs the framework of General Relativity, and in particular the inclusion of the gravitational field, in order to be able to put the equations of motions into a form that holds equally in *all* reference systems. In this fashion the logically somewhat dissatisfying distinction between 'real' and 'inertial' forces can also be overcome. In SR, however, this distinction as well as the special status given to inertial systems is maintained.

2.3 Is the relativity principle valid in electrodynamics?

If the electromagnetic field is to be understood as a function of state of an ether, then the ether must be present wherever one detects the electromagnetic field, in particular in the interior of ordinary 'ponderable' matter. There it will suffer some interaction with that matter. One may, e.g. envisage a difference of the ether density inside and outside matter, which might explain the difference in properties of the electromagnetic field, like, e.g. the various propagation speeds of electromagnetic waves including light. For example, it is known in the theory of elasticity that the propagation speed is inversely proportional to the square-root of the material's density. By analogy this might mean that the diminished speed of light within materials is due to a higher density of the ether (somewhat counter-intuitive to the naive expectation that the ether gets displaced by ordinary matter). Such a theory was indeed attempted quite early by the French physicist Augustin Jean Fresnel (1788–1827).

Quite generally, the speed of light (more precisely, its phase velocity; see Sect. 5.3) in materials is written as

$$c_m = \frac{c}{n}, \tag{2.8}$$

where c is the speed outside any ponderable matter (i.e. 'in vacuum'), and where n denotes the 'index of refraction' for the material in question. To be precise, one has to add that c_m as well as c are meant relative to the local rest frame of the ether. (Here and below we shall understand the word 'motion' as applied to the ether always in some coarse-grained sense that averages over the small vortices which according to Maxwell's ideas are the causes of certain electromagnetic fields.) If one moves relative to the ether one expects the 'ether-wind' to carry the wave preferably into the direction in which the wind blows. In other words, the speed of light should depend on the direction, i.e. be anisotropic. But this sounds as if the principle of relativity cannot be extended to electromagnetism, because measuring the speed of light in all directions would allow one to deduce one's state of motion relative to the ether. Is it then true that the principle of relativity is violated in electromagnetism? An answer to this question can only be given by the theory of *electrodynamics of moving bodies.*

Let us recall Salviati's (i.e. Galileo's) description of the relativity principle in mechanics. As we stressed in our discussion there, it was essential that *all* components of the physical system participated alike in the uniform translational motion of the ship. This is why going 'below decks' was essential in order to prevent the air from flowing through, which would otherwise give away the ship's state of motion. In electrodynamics the rôle of the air is taken up by the ether. Would the ether just waft through ordinary matter, so that 'going below decks' is no option in electrodynamics? If yes, there could be no relativity principle in electromagnetism. Hence the all-important question is, whether and how the ether is dragged along by the motion of matter.

2.4 Experiments, contradictions, and consequences

During the last three centuries many experiments have been performed in connection with the questions just raised. Here we wish to discuss those classic ones which are directly concerned with light propagation. These were complemented by others which looked more closely into the detailed behaviour of electric and magnetic fields in moving media and which are equally important. The reason why we restrict attention to optical experiments is that they can be described in sufficient detail without entering

22   *Historical developments*

the electromagnetic theory proper. Hence it should be kept in mind that more involved investigations support the arguments given and conclusions drawn here.

2.4.1 Aberration

First we turn to the phenomenon of aberration, which we illustrate in **Fig. 2.5**. If one observes a fixed star by a telescope, the light emitted by the star has to pass both the eyepiece at the rear end and the lens at the front end of the telescope. In case the telescope and the star are both at rest relative to the ether, the situation is as depicted in the left picture of **Fig. 2.5**, where eyepiece, lens and star form a straight line. In contrast, the right pictures depict a situation where the telescope now moves to the right and perpendicular to the original line of vision relative to the star. The star is taken to be at rest relative to the ether. If we assume that there is no dragging of the ether by the telescope, then one has to tilt the telescope into the direction of motion in order for the light ray to

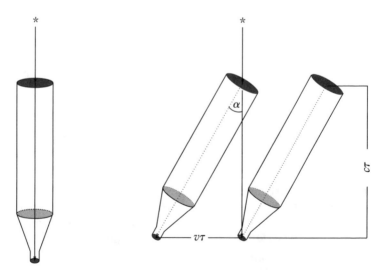

Fig. 2.5 The phenomenon of aberration.

Experiments, contradictions, and consequences 23

pass the lens *and* the eyepiece. The first picture on the right shows the telescope at the time where a particular light phase hits the lens, the second at the time where this phase just escapes the eyepiece. Within this time interval, τ, in which the starlight travelled from the lens to the eyepiece, the telescope has moved a distance $v\tau$ to the right. Hence we must tilt the telescope by a definite angle α in order for the lens and the eyepiece to lie on the same vertical light ray when considered a time interval τ apart. Since light travels along the vertical rays with speed c, the vertical separation between lens and eyepiece equals $c\tau$. Hence we get the following relation between the tangent of the tilting angle α and the velocities v and c:

$$\tan \alpha = \frac{v}{c}. \tag{2.9}$$

This relation was already used in 1728 by the English astronomer James Bradley (1673–1762), who had just discovered the effect of aberration three years earlier, in order to determine the speed of light c by measuring aberration angles. He observed a fixed star in a direction almost perpendicular to the ecliptic (the plane defined by the orbit of the Earth) so that the line of sight would be nearly perpendicular to the velocity of the Earth. The Earth's velocity, v, on its nearly circular orbit is given by 2π AU/year, where AU is the so-called Astronomical Unit, which denotes the (averaged) distance between the Earth and the Sun. For example, taking the (rounded) modern value AU = 150 million kilometres, we obtain v close to 30 kilometres per second. On the other hand, the aberration angle measured by Bradley is about 20″, where ″ denotes arc-seconds, corresponding to 10^{-4} radians. The resolution achieved by Bradley was about 1″. Since for small angles we can approximate the tangent by its argument measured in radians, we have from (2.9) that $c = v/\alpha$. Putting in the numbers we get $c = 3 \cdot 10^8$ m/s. Bradley obtained a value remarkably close to this (see below).

During the course of a year the axis of a telescope looking at a fixed star describes a cone of opening angle 20″ (measured from the axis to the cone). It is important to realize that this angle is much larger than the largest annual parallax, which is $0.76''$ for our closest neighbour (Proxima Centauri, distance 4.22 light years) and hence outside the resolution of Bradley's instruments. Also, aberration differs from parallax in that the latter decays with distance. In fact, Bradley observed the star γ Draconis (a red giant and the brightest star in the constellation of the Dragon, one of the four stars forming its head) whose distance to us is about 100–130 light years. He measured the apparent annual variation in position and from that the aberration angle. Then he used (2.9) to obtain the remarkably accurate value of $1/10186$ for v/c, which differs by slightly more than one per cent from the modern value $1/10060$. Using the most accurate value for AU available to him to calculate v, he finally determined c with much improved accuracy as compared to that obtained by Rømer 50 years earlier (cf. Sect. 5.1). In particular Bradley's method gave an independent proof for the finiteness of the speed of light which finally convinced the remaining (if any) critics. Also, it can be viewed as the first direct evidence for the motion of the Earth, since the first successful and accurate measurements of star parallax were performed 110 years later by the German astronomer and mathematician Friedrich Wilhelm Bessel (1784–1846) in Königsberg (then Eastern Prussia).

Central to the question of a possible ether drag is not so much Bradley's determination of c, but the fact that we can test (2.9), given the value for c from *independent* measurements. The derivation of (2.9) makes it clear that this relation is only valid if the ether within the telescope is not significantly dragged along. If there was a complete drag there would certainly be no aberration. Since the interior of the telescope consists mainly of air-filled space, except for the lenses, it was thought that filling the telescope with water might enhance the drag and hence diminish the aberration. This experiment was indeed performed in 1871 by the English astronomer George Biddell Airy (1801–1892), but no such influence of water was seen.

2.4.2 Fizeau's experiment

Another experiment of great historical interest and some ingenuity is that performed in 1851 by the French physicist Armand Hippolyte Fizeau (1819–1896). **Fig. 2.6** is a sketch of this experiment in a slightly modified form. A source L sends light to a half-silvered mirror T which separates the beam into two components. One component consists of the light reflected by T. It gets reflected by the mirrors S_1, S_2, and S_3 in chronological order before it returns back to T, where parts of it get reflected to the observer B. (The part that by passing T runs back to L does not interest us.) The other component consists of the light that passes through T and then travels the rectangle counterclockwise, i.e. reflections take place at S_3, S_2, and S_1 in chronological order until the light finally passes T to reach the observer B. (Again the light reflected at T to run back to L does not interest us.) In B the observer has some suitable arrangement to measure the interference pattern produced by the two superposed components. On the horizontal parts of its route the light passes along the interior of a pipe whose walls are made of glass and which contains a liquid of refractive

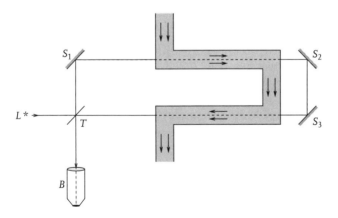

Fig. 2.6 Optical paths in Fizeau's experiment.

index n (indicated in grey in **Fig. 2.6**). This liquid can be set into flowing motion as indicated by the arrows.

By the geometry (U-shape) of the pipe the light component that travels round the rectangle in a clockwise fashion will, within the pipe, always run parallel to the flow direction of the liquid. For the counterclockwise travelling component the light travels antiparallel to the flow of the liquid. If the flowing liquid would cause an ether drag it would accelerate the first (clockwise travelling) component in the pipe and decelerate the second. The idea now is to first observe the interference fringes when the liquid is at rest. Then one sets the liquid in motion and observes whether and how the interference fringes shift. Such a shift should exist if there is an ether drag, since then the first component takes less and the second more time to complete the circuit. Fizeau used water as the filling for the pipes and indeed found a characteristic shift of the interference fringes which he interpreted as due to the drag-modified propagation speeds of light in the tube. If c/n denotes the speed of light in the liquid and relative to it, and v the speed of the flow relative to the laboratory system, then one may express the speed of light in the liquid relative to the laboratory as

$$c' = \frac{c}{n} + v\varphi. \tag{2.10}$$

Here φ denotes the so-called 'drag coefficient'. It parametrizes the degree to which the liquid drags the ether. For $\varphi = 1$ the drag is complete and for $\varphi = 0$ there is no drag at all. Fizeau could represent his results by the following formula, which had already been suggested by Fresnel on the basis of his (wildly speculative) ether theory:

$$\varphi = 1 - \frac{1}{n^2}. \tag{2.11}$$

Fizeau used $n = 1.33$ corresponding to water. Much later, 1914–19, the Dutch physicist Pieter Zeeman (1865–1943) carefully repeated Fizeau's experiments and also replaced the streaming liquid by solid bodies made of glass or quartz in order to attain higher values of n. He found full agreement with (2.11) if dispersion (see below) is negligible, as has been assumed so far. He also verified an improved version of (2.11) that takes into account dispersion.

In connection with (2.11) there are three points worth contemplating. First, that there is a drag at all, i.e. that $\varphi \neq 0$. Second, that the drag is not complete, i.e. that $\varphi \neq 1$. Third, that the drag depends on the refractive index (and nothing else). For gases n is very close to one, like $n = 1.00014$ for air, whereas for solids it lies between 1·5 for glass and 2·4 for diamond. Having said this, we must not suppress the fact that the n-dependence of the ether drag causes severe theoretical difficulties. This has to do with the existence of 'dispersion', i.e. the dependence of n on the light's wavelength. Hence, strictly speaking, the refractive index is not a number uniquely associated with the material in question, but rather a function of the wavelength. The above cited numbers are then reference values evaluated at some fixed wavelength, here within the visible spectrum, where the variation of n upon the wavelength is small. Still such dependencies lead to familiar visible effects, like the rainbow, where dispersion is responsible for the decomposition of white light into its spectral colours. Applied to relation (2.11) this means that the degree of drag depends on the colour, which seems to contradict the whole idea that there is only *one* ether being dragged and that this ether carries the light waves of all frequencies.

Despite this and other objections, the incompleteness of the ether drag had for a long time been seen as proof for the existence of the ether. How otherwise should one understand the apparent fact that light can travel in one and the same material with different velocities: with velocity c/n if the material is at rest with respect to the ether and with velocity

$$c' - v = \frac{c}{n} - \frac{v}{n^2}, \qquad (2.12)$$

if it is moving in the direction of the light ray with velocity v relative to the ether? Does this not clearly show the effect of an 'ether-headwind'? It is admitted that the velocity (2.12) of light relative to the streaming liquid has not been directly measured in Fizeau's experiment, but only the velocity relative to the laboratory. However, a simple application of the law for the addition of velocities, (2.5), seems to unambiguously lead from (2.10) to (2.12). The theory of SR will later unmask this as erroneous. But for the time being this was regarded as indubitable. Hence it should be possible to measure directly variations in the speed of light depending on the state of motion relative to the ether.

2.4.3 The Michelson–Morley experiment

In 1879 Maxwell made an interesting and fairly obvious suggestion to directly measure the velocity of the solar system relative to the ether. He pointed out that Rømer's method (see Sect. 5.1) measured the speed of light along the Earth's orbital diameter in a particular direction, namely that pointing from Jupiter to the Sun. Since the orbital period of Jupiter is about 12 years, Maxwell suggested comparing many Rømer-like measurements performed over an extended period that included at least half a Jupiter revolution, i.e. six years. These would yield the speed of light in different directions, including diametrically opposite ones. (Here one assumes that the direction of the ether wind relative to the solar system is approximately constant within this six years.) A non-vanishing ether flow should then result in a periodic modulation of the measured speed. As function of time the period would be 12 years.

The interesting aspect of this suggestion is that it is relatively independent of any assumption concerning the possible drag of the ether caused by the Earth's atmosphere or any parts of the experimental arrangements. This is because the by far dominant part of the light path from Jupiter to Earth lies outside those structures. But unfortunately this experiment must fail for reasons that lie in an inherent inaccuracy of Rømer's method and the actual orders of magnitude involved (which were partly unknown to Maxwell). Let us explain this: Today we know that typical relative velocities of stars in our galaxy are given by a few hundred kilometres per second. For example, our solar system orbits our galactic center at a speed of approximately 220 km/s. Hence we would naturally not expect the speed of the solar system relative to the cosmic ether to be significantly less. A variation of the speed of light by that amount results in timing variations of signals of at most two seconds over the distance from Jupiter to us. But the 'signals' used by Rømer's method are the eclipses of Io (the innermost Galilean moon of Jupiter) caused by Io entering Jupiter's shadow. This process of entering the shadow is not a sudden event but a process that observationally can be resolved in time with an accuracy of about a minute. Hence we are almost two orders of magnitude above the required timing resolution. Even if one achieved the extremely optimistic resolution for the Io eclipse of 10 seconds, one could still not determine ether velocities below 500 km/s, which is certainly not sufficient.

The first real experimental breakthrough with respect to the required accuracy was achieved in 1887 by the American experimentalists Albert Abraham Michelson (1852–1931) and Edward Morley (1836–1923) [4], who repeated an experiment that Michelson had already performed with considerably lesser accuracy in 1881 at the Astrophysical Observatory at Potsdam near Berlin. First, by mistake, Michelson overestimated the expected effect by a factor of two. Correcting this, his findings were just of the same order as the experimental errors, so that his experiment remained inconclusive. This was substantially improved on by the second version of 1887. See [5] for more on the interesting history concerning these experiments.

Figure 2.7 depicts the basic idea of the experimental setup of Michelson and Morley, which like in Fizeau's experiment, is based on an interferometer. Here, too, the source L sends a beam of light on to a half-silvered mirror T which decomposes the incoming beam into two components of approximately equal intensity. The component which is depicted horizontally in **Fig.** 2.7 travels the distance l_1 to a mirror S_1 where it is reflected back to T and there finally (partly) reflected to the observer B. (The part passing T and reaching L does not interest us.) The vertical component travels the distance l_2 until the mirror S_2 reflects it back to T where parts of

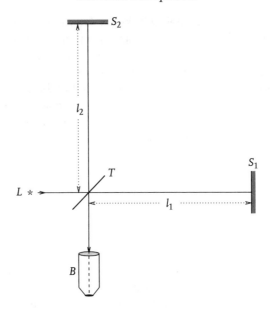

Fig. 2.7 The experiment of Michelson and Morley.

it pass T and reach the observer B. (Again, the part being reflected back to L does not interest us.) At B an appropriate arrangement allows to observe the interference patterns of the two superposed components. The optical setup is mounted on a thick stony base that floats in a basin filled with liquid mercury, so as to be able to rigidly rotate the whole optical component with the smallest possible mechanical disturbances.

The whole setup is firmly attached to the surface of the Earth and therefore fully participates in its motion, which is essentially composed of its diurnal spinning revolutions and its annual orbital revolutions around the Sun. According to Fizeau's result (2.11) one would not expect any measurable ether drag caused by the Earth's atmosphere ($n \approx 1$ for gases). Hence one would expect an ether wind to blow just above the Earth's surface. Its speed should not be less than the orbital speed of the Earth relative to the Sun, at least not on average over a year's time. The Michelson–Morley experiment

Experiments, contradictions, and consequences 31

now tests whether the difference between the travel times (forth *and* back) along the two arms depends on the orientation of the apparatus. According to the ether theory a definite dependence should be expected, as the following discussion shows.

We describe the process from the rest system of the ether. For simplicity we also assume the velocity relative to the ether, v, to be in the horizontal direction, pointing from T to S_1. In other words, the 'ether wind' blows from the right at speed v. Hence light travels from T to S_1 at speed $c - v$ and backward at speed $c + v$ relative to the apparatus. The travel forth and back in the horizontal direction takes time

$$T_1 = \frac{l_1}{c-v} + \frac{l_1}{c+v} = \frac{2l_1}{c}\gamma^2, \tag{2.13}$$

where here and in the sequel we use the abbreviations

$$\beta = \frac{v}{c} \quad \text{and} \quad \gamma = \frac{1}{\sqrt{1-v^2/c^2}}. \tag{2.14}$$

γ is called the gamma-factor belonging to the velocity in question. Sometimes we will explicitly indicate the velocity parameter on which gamma depends and write, e.g. $\gamma(v)$. This will prevent confusion if more than one velocity parameter is involved in the discussion.

In order to calculate the travel time in the vertical direction, i.e. from T to S_2 and back, we take a look at **Fig. 2.8**. The couple of mirrors consisting of T and S_2 moves in a horizontal direction as the light travels from T to S_2. In **Fig. 2.8** T, S_2 denotes that pair at the time the light hits T; T', S_2' at the time it hits S_2; and finally T'', S_2'' at the time it returns to T. By τ we denote the time interval the light needs to travel from T to S_2. In this time interval S_2 has moved by an amount $v\tau$. Likewise, during the travel from S_2 back to T, the half-silvered mirror T has itself moved by the same amount $v\tau$ from T' to T''. Since $TT'S_2'$ and $S_2'T'T''$ are right angled triangles, the Pythagorean theorem implies for the given length that $c^2\tau^2 = l_2^2 + v^2\tau^2$, which can be easily solved for τ,

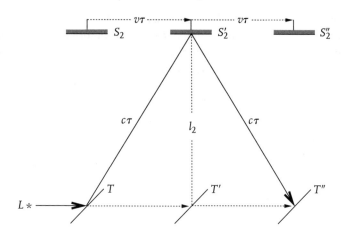

Fig. 2.8 Optical path in the transverse direction, seen from the rest system of the ether.

which is just half the vertical travel time T_2. Hence we have

$$T_2 = 2\tau = \frac{2l_2}{c}\gamma. \qquad (2.15)$$

Note that this differs from (2.13) only in that γ (cf. 2.14) enters linearly here but quadratically there.

Let us specialize to the case where the source emits monochromatic light of frequency ν. A fixed phase of light that travelled along the horizontal arm is seen at B with a delay $T_1 - T_2$ as compared to the same phase that travelled along the vertical arm. In other words, it arrives with a delay of

$$N = \nu(T_1 - T_2) \qquad (2.16)$$

number of phases. This is seen as a fixed interference pattern at B. Turning the interferometer by 90 degrees in, say, a clockwise direction exchanges the rôles of the horizontal and vertical arm respectively. The ether wind is now blowing in the direction from S_2 to T whereas the line TS_1 is now perpendicular to the ether wind. By a simple repetition of the argument above we obtain the travel times T'_1 and T'_2 along the arms TS_1 and TS_2, respectively, after rotation:

$$T'_1 = \frac{2l_1}{c}\gamma \quad \text{and} \quad T'_2 = \frac{2l_2}{c}\gamma^2. \tag{2.17}$$

The number N' of phases by which the light that travels over S_1 is delayed with respect to the light that travels over S_2 is now given by

$$N' = \nu(T'_1 - T'_2). \tag{2.18}$$

Hence the difference $N - N'$ just corresponds to the number of interference fringes seen shifting at B in the process of a 90 degree rotation of the apparatus. If we express the frequency ν by the wavelength $\lambda = c/\nu$, we obtain

$$\Delta N = N - N' = 2\frac{l_1 + l_2}{\lambda}\gamma(\gamma - 1) \approx \frac{l_1 + l_2}{\lambda}\beta^2, \tag{2.19}$$

where the second expression on the right hand side is valid approximately for small values of β (cf. 2.14), that is, for velocities v small compared to the velocity of light. Note that β enters quadratically which means that for small β this effect is strongly suppressed as compared to effects linear in β, like aberration (2.9). Setting $v = 30$ km/s for the orbital motion of the Earth and $c = 3 \cdot 10^5$ km/s for the velocity of light one has $\beta = 10^{-4}$ and $\beta^2 = 10^{-8}$. This means that quadratic effects are smaller by a factor of ten thousand as compared to linear ones. Michelson and Morley used equal arm lengths of effectively 11 metres optical length (through repetitive reflections they multiplied the geometric length) and light of wavelength 5900 Å (Å = Ångström = 10^{-10} m) which is of a yellow colour. Hence they expected to see a shift in the interference fringes of $\Delta N = 0.37$. Their resolution was high enough to measure $\Delta N = 0.01$ so that the expected effect was almost 40 times larger than this.

The surprising result was that, within this accuracy, *no* effect was seen at all. For the ether theory this could only mean that the velocity of the ether wind was much smaller—at least 40 times—than the orbital velocity of the Earth. But even if by sheer chance the Earth was at rest relative to the ether during the days the experiment was performed, one would just have to wait for 6 months and repeat it with an oppositely directed orbital velocity of the Earth. The relative velocity between ether and Earth should then be around 60 km/s. The experiment was indeed repeated at different times of the year but the result turned out to be always null.

Daring explanations were offered to explain these null results. It was, e.g. argued that perhaps the lower layers of the atmosphere could nevertheless show some dragging effect (contrary to Fizeau's result) or, more seriously, that such a dragging could be provided by the material of the heavy walls of the laboratory, by which the experiment was protected from any disturbances from the outside world. Anticipating the historical development, we mention that Dayton Miller (1866–1941), a former collaborator of Morley's, repeated the experiment in 1921. He positioned his interferometer in a hardly shielded little hut on top of Mount Wilson, so as to enable the ether wind to waft through as unhindered as possible. And indeed, he reported a positive (non-null) result, against the predictions of SR, the theory of which had then already existed for 16 years! Einstein, who was in the USA at this time, commented on this with his now famous 'subtle is the Lord, but he is not malicious'. And he was quite right. Subsequent runs of the experiment could not confirm any positive result and it is now believed that Miller's first result was just based on an experimental error. Max Born (1882–1970), a lifelong friend of Einstein's and master of theoretical optics and SR, visited Miller's laboratory in 1925/26 and 'was shocked by the shaky arrangements', as reported by his wife. In his own words:

I found it [the experimental arrangement] shaky and unreliable; the smallest movement with one's hands or a cough made the interference fringes so fidgety that there was no way to read off their position. After that I didn't believe a word of his experiments. From my stay in Chicago in the year 1912 I knew of the reliability of Michelson's own apparatuses and the accuracy of his measurements.

As a consequence of the (intended) poor protection from outside influences, the accuracy achieved by Miller was indeed much worse than that of the original 1887 Michelson–Morley experiment. A tenfold improvement in accuracy over the original experiment was first achieved in 1930 by Georg Joos (1894–1959) in Jena (Germany). He too found no effect of an ether wind. Modern experiments improve on this by many orders of magnitude; see Sect. 5.6.

2.4.4 The FitzGerald–Lorentz deformation hypothesis

Hardly two years after the experiment of Michelson and Morley a half-page note appeared in the American science journal *Science* entitled 'The Ether and the Earth's Atmosphere'. The author was the Irish physicist George Francis FitzGerald (1851–1901), who proposed the following seemingly outrageous hypothesis to explain the null result of the Michelson–Morley experiment: any motion of a body through the ether universally affects its geometric dimensions. Here 'universally' means that it affects all materials alike, independent of their physical or chemical state, just depending on the magnitude of their velocity relative to the ether. This suggestion went unnoticed until three years later when, in 1892, the Dutch physicist Hendrik Antoon Lorentz (1853–1928), apparently independently, proposed a concretization of the same idea. In fact, such an hypothesis was not as outrageous as it first might have appeared if one assumed an atomistic viewpoint, in which the constitution of a solid was exclusively determined by the electrostatic forces between its elementary constituents (atoms, molecules). In 1888 Oliver Heaviside (1850–1925) deduced from Maxwell's equations that the electric field of a spherical charge distribution in motion is 'squashed' in the direction of motion as compared to the field of a charge at rest, see **Fig. 3.15**. Recall that in those days it was assumed that Maxwell's equations refer to the ether's rest system so that 'moving' and 'at rest' make good sense. According to this one could conjecture that all bodies shrink their geometric size in the direction of their velocity relative to the ether by the same factor that describes the squashing of the electric field. It was of course not known whether all forces in materials were finally reducible to electromagnetic ones. But that was at least a natural hypothesis which lent a certain plausibility to what originally merely seemed to be a purely ad hoc assumption by FitzGerald and Lorentz. A more detailed and interesting account of this story is given in [6].

To demonstrate that the FitzGerald–Lorentz hypothesis does indeed explain the null result of the Michelson–Morley experiment, we first assume the whole arrangement of **Fig. 2.7** was at rest relative to the ether. In that state the arm lengths are denoted by l_1^0 and l_2^0. We now assume that if the whole setup is put into a state of motion, all lengths in the direction of motion are scaled by a factor of A and all lengths transversal to it by a factor of B. A and B just depend on the magnitude v of the relative velocity with respect to the ether, and nothing else. This means that all materials in all states are affected alike. That a moving body 'changes length' is always meant relative to a yardstick which itself is at rest relative to the ether. If the moving object is measured with an equally moving yardstick, no such change is seen, since, by hypothesis, both are affected alike. In the following all lengths are therefore understood to be measured with yardsticks at rest relative to the ether. If l_1 and l_2 denote the arm lengths in the state of motion before the apparatus is rotated, and l_1' and l_2' the lengths of the physically same arms after rotation, we should have, according to the hypothesis,

$$A = \frac{l_1}{l_1^0} = \frac{l_2'}{l_2^0} \quad \text{and} \quad B = \frac{l_2}{l_2^0} = \frac{l_1'}{l_1^0}. \tag{2.20}$$

This is now used in expressions (2.13) for T_1 and (2.15) for T_2 to re-express l_1 and l_2 by Al_1^0 and Bl_2^0, respectively. The same is done in expressions (2.17) for T_1' and T_2', where one first has to rename l_1 and l_2 as l_1' and l_2' according to our notation used here. Having done that, one obtains instead of (2.19) for the number of shifted interference fringes during rotation:

$$\Delta N = 2 \frac{l_1^0 + l_2^0}{\lambda} \gamma (\gamma A - B). \tag{2.21}$$

This shows that a null result, i.e. $\Delta N = 0$, of the Michelson–Morley experiment can be explained by further imposing the following relation between the longitudinal deformation factor A and the transversal deformation factor B:

$$A\gamma = B. \tag{2.22}$$

The last expression in brackets on the right hand side of (2.21) then vanishes. In particular, it would be sufficient—though not necessary—to take $B = 1$ and $A = 1/\gamma$. This corresponds to a longitudinal contraction and no transversal deformation. This is the case that emerged from electrodynamics on account of Heaviside's calculations mentioned above. We will see that SR too will predict precisely that longitudinal contraction, however without involving any ether theory. But we stress again that on account of the Michelson–Morley experiment alone only the quotient of A and B is fixed. To fix the actual values of A and B one has to involve two more experiments, like e.g. the Kennedy–Thorndike and Ives–Stilwell experiments. These will be discussed in Sects. 5.4 and 5.5, respectively.

·3·
Foundations of Special Relativity

The foregoing discussion should have made clear the state of tension in which physics found itself at the end of the 19th century. This was already felt by the young and bright Einstein. Shortly before his death he remembered his short but happy year (1895–1896) as a pupil of the school in the small city of Aarau (Switzerland), in whose liberal atmosphere the 'saucy Swabian' (according to a classmate) felt much more at home than in the authoritarian Gymnasium at Munich (Germany).

During that year in Aarau the following question came to my mind: If one follows a light wave at the speed of light, one would be confronted with a time-independent wave field. But such a thing does not seem to exist. This was the first childish gedanken experiment in connection with Special Relativity.

Maxwell's theory indeed predicted electromagnetic waves, but always at an invariant propagation speed outside matter. So, as long as one believed an ether to exist, one had to assume Maxwell's equations to be valid exclusively in the rest system of the ether. For if they were also valid in other inertial systems, for example that one that chased after a light wave at the speed of light, then, as Einstein remarked, the moving observer should see a standing light wave, which must be a solution of Maxwell's equations. But the latter is definitely not the case.

Actually, we already know that our intuition is very probably playing a trick on us here, because we implicitly implied the validity of the classical law (2.5) for addition of velocities when we assumed that for the moving observer the light wave was standing. However, the experiments of Fizeau and Michelson–Morley strongly suggest

Foundations of Special Relativity 39

that processes concerning the propagation of light do not obey this law. (Note that the experiments discussed in Sect. 5.2 post-date SR.) This was certainly known to Einstein at the time he conceived SR, though it is likely that he knew of the Michelson–Morley experiment only indirectly, e.g. through Lorentz's then famous monograph of 1895 [7], entitled 'Inquiry into a Theory of Electrical and Optical Phenomena in Moving Bodies'. Among other experiments, the Michelson–Morley experiment is discussed in the last section of this book, which is entitled 'Experiments, the results of which cannot be explained offhand'. Here Lorentz discusses the deformation hypothesis (cf. Sect. 2.4.4) that enforces such an explanation. Lorentz also points out that, assuming all molecular forces to be of electromagnetic origin, the deformation would merely be a longitudinal contraction without transversal change, but at the same time admits that this is (in 1895) a physically unwarranted hypothesis. An interesting portrayal of Einstein's own recollections, concerning the impact that the Michelson–Morley experiment had on the formation of SR, is given in [8].

Characteristic of Einstein's scientific thinking was his finely developed sensitivity for conceptual imbalances. For him, the eventual elimination of such difficulties were mandatory, ranking no less in priority than the elimination of plain experimental contradictions. In fact, in his original SR paper [9], Einstein did not mention or cite a single experiment explicitly. Only an incidental and unspecific remark concerning 'experiments' is made (see the quotation below). Instead, Einstein devotes his entire opening paragraph to the discussion of the seemingly harmless phenomenon of electromagnetic induction, which is well known from electrical engineering—being the basis of any electric engine—and which is still today one of the favourite themes in high-school physics. To demonstrate the point Einstein emphasizes, take a look at the U-shaped (with open end to the right) piece of wire, depicted in **Fig. 3.1**. Perpendicular to the plane of the paper is a magnetic field that pierces the plane in the region marked by the symbols ⊗ in an upward direction. If the conductor moves relative to the magnetic field in the direction of the arrow, a voltage is induced

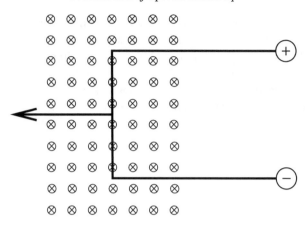

Fig. 3.1 Induction of a voltage in a conductor (wire) that moves relative to a magnetic field.

at its ends whose signs are as indicated. Einstein now points out that, according to the then current interpretation of electromagnetic theory, the explanation of this phenomenon is strongly dependent on whether one regards the magnetic field or the conductor as the moving part, even though the phenomenon itself is perfectly symmetric. Note that in an ether theory such a distinction makes good sense, since there is the ether's rest frame with respect to which 'motion' can be defined. If the magnetic field is moving, i.e. time dependent, Maxwell's equations predict an electric field being induced inside and outside the conductor, which is so directed that the electrons move to the lower end, thereby causing an electron shortage at the upper end (and hence a positive voltage) and an electron abundance at the lower end (and hence a negative voltage). Conversely, if the magnetic field is at rest, i.e. time independent, Maxwell's equations predict that there is no electric field, neither inside nor outside the conductor. Rather, the electrons of the conductor material will now feel a force due to their motion relative to the magnetic field. This force is called the 'Lorentz force'. It points in a direction which is perpendicular to the magnetic field and at the same time perpendicular to the velocity of the charge relative to the magnetic field, thereby obeying the right-hand rule. Hence, in our case, it points in an axial direction along the wire, thereby causing the electrons to move to the lower end. As a result, the same voltage as before is obtained. After having outlined this, Einstein ends his introductory section as follows [9]:

Examples of this sort, together with the unsuccessful attempts to discover any motion of the Earth relative to the 'light medium', suggest that the phenomena of electrodynamics as well as of mechanics possess no properties corresponding to the idea of absolute rest. They suggest rather that [...] the same laws of electrodynamics and optics will be valid for all frames of reference for which the equations of motion hold good. We will raise this conjecture (the purport of which will hereafter be called the 'Principle of Relativity') to the status of a postulate, and also introduce another postulate, which is only apparently irreconcilable with the former, namely, that light is propagating in empty space with a definite velocity c which is independent of the state of motion of the emitting body. [...] The Introduction of a 'luminiferous ether' will prove to be superfluous inasmuch as the view here to be developed will not require an 'absolutely stationary space' provided with special properties, nor assign a velocity-vector to a point in empty space in which electromagnetic processes take place.

The theory to be developed is based—like all electrodynamics—on the kinematics of the rigid body, since the assertions of any such theory concern relationships between rigid bodies (coordinate systems), clocks, and electrodynamic processes. Insufficient consideration of this circumstance lies at the root of the difficulties which the electrodynamics of moving bodies presently encounters.

On the subsequent pages Einstein shows that a precise redefinition of kinematical quantities is sufficient to establish the principle of relativity also in electrodynamics and at the same time reconcile it with the universality of the speed of light! Thereby the ether becomes conceptually dispensable and, in that sense, physically abolished.

3.1 The notion of simultaneity

We measure the length of a body at rest by comparing it with a yardstick. More precisely, we put equidistant marks on the yardstick and read off first one, and then the other bounding mark. The number of marks in between, plus one, is then the length of the body in the chosen units. But how would we proceed if the body was moving relative to the yardstick? An obvious way would be to read off the marks *simultaneously* from the yardstick at the moment the body passes it. But the crucial point here is that the two events where the two marks are read off are spatially separated. This means that one needs to employ some definition of simultaneity of spatially separated events in order to give this procedure a well defined meaning.

It is important to realize that even the most elementary statements concerning motion refer to simultaneous events. Saying that a train arrives at time t at a particular train station, x, means that the event of arrival at x and the event where the station clock strikes t are simultaneous. For this to be a meaningful statement the clock and train need to be at a close distance. If the clock closest to the train is at a distance d, we see its hands at the time they were when the light that we now see left the clock; that is, a time d/c earlier than now. In order to not miss the train this delay should not exceed, say, one minute. This implies that the clock should be within a distance of 18 million kilometres. Whereas this clearly sounds like a ridiculous constraint as far as train journeys are concerned, it does become important for much larger velocities and smaller time scales. If the duration of a 'train' stop is of the order of nanoseconds (10^{-9} s) the maximal distance a clock may be located at is already down at 30 centimetres. And, finally, it is clear that the timing of processes that move close to the velocity of light require strictly local definitions of time, since here the retardation times d/c are just of the same order of magnitude as the typical durations one wishes to measure.

Let now K be an inertial reference system. Attaching identical copies of a clock to each point of K does not yet define a notion of 'time'. This is simply because the clocks need to be synchronized in order to speak of 'the same time' at different locations. Only after synchronization does it make sense to assert that a clock at position A and another clock at position B show the same time t, if their hands are in identical configurations showing t. An obvious and conceptually simple procedure for synchronization is to move a 'transport clock' successively to each point in space and, while being at the same point, synchronize it with the local clock. The pointwise synchronization just consists in putting the hands of both clocks simultaneously (at the same point in space) to the same position. The transport itself has to be done sufficiently carefully in order not to upset the clock's rate through mechanical disturbances. Clearly, for widely separated clocks, this procedure is hardly feasible. Einstein therefore suggests an alternative method which avoids the transport of clocks. Rather, the locally fixed clocks are synchronized by the exchange of light signals. More precisely, the synchronization of clock B with clock A then proceeds as follows. A light signal is sent from A to B, where it will be instantaneously reflected back to A. Let $t_A^{(1)}$ and $t_A^{(2)}$ denote the simultaneous readings of A with the events of emission and re-absorption, and t_B the reading of B simultaneous with the event of receiving the light from A. Now, B is said to be synchronized with A if and only if

$$t_B = \frac{t_A^{(1)} + t_A^{(2)}}{2}. \tag{3.1}$$

This is clearly the same as saying that the travel time from A to B is the same as the travel time back from B to A. Therefore, another equivalent condition is this: consider two light signals, one sent at t_A from A to B, the other at t_B from B to A. The clocks at A and B are synchronized, if and only if the light signals simultaneously pass the midpoint of the segment $\overline{AB}$.

It is of central importance to realize that this room for convention indeed exists, i.e. that there are no facts of experience, logically independent of clock synchronization, which further restrict this freedom. The only synchronization-independent statement that we can make about the velocity of light is to measure, by a *single* clock, the time it needs for a round trip. If d is the distance between A and B, then the *mean* velocity of the light on its way from A to B and back is given by $2d/(t_A^{(2)} - t_A^{(1)})$. The one-way velocities from A to

B, or B to A, are not defined without stipulating a synchronization procedure.

The meaning of a synchronization procedure is precisely to allow the definition of simultaneity of spatially distant events, by reducing it to assertions concerning the simultaneity of equilocal events:

Definition of simultaneity *Two events at spatially separated locations A and B are called simultaneous, if the locally simultaneous clock readings of synchronized clocks at A and B are identical.*

In this way, Einstein's synchronization procedure leads to a notion of simultaneity of events, i.e. space-time points, that obeys the following laws:

(1) Every event is simultaneous to itself.
(2) If p is simultaneous to q then q is simultaneous to p.
(3) If p is simultaneous to q and q is simultaneous to r, then p is simultaneous to r.

Quite generally, a relation between pairs of points taken from some arbitrary set is called an 'equivalence relation' if it satisfies (1)–(3). Conditions (1)—called reflexivity—and (2)—called symmetry—are natural for any notion of simultaneity, though one may also envisage non-symmetric generalizations. Condition (3)—called transitivity—is necessary in order to generalize the notion of pairwise simultaneity to the notion of mutual simultaneity of sets containing more than two events. In particular, the following is quite easily seen to hold true for any simultaneity relation that satisfies (1)–(3): Let R_p and R_q be the sets of events simultaneous to the events p and q, respectively, i.e. their 'equivalence classes'. Then R_p and R_q are either disjoint (have no point in common) or identical. In other words, two simultaneity classes cannot intersect in a proper subset. Therefore, Einstein's definition of simultaneity partitions space-time into mutually disjoint sets, each containing mutually simultaneous events. The converse is also obviously true, namely that any partitioning into mutually disjoint sets defines an equivalence relation. These conditions therefore should be expected from any workable definition of simultaneity, though relaxations may be considered, in particular concerning transitivity.

The notion of simultaneity

Now, the all-important point is this: The synchronization procedure discussed so far applies to clocks at rest in a particular inertial system of reference K. Hence the resulting notion of simultaneity is tied to K. Suppose we repeat the synchronization procedure with clocks at rest in a different system, K', where K' moves uniformly relative to K. Would this give rise to a different notion of simultaneity on space-time? More precisely: would the two equivalence classes R_p and R'_p of points simultaneous to the same event p, one time determined according to the K-simultaneity and the other according to the K'-simultaneity, be different sets? The answer is in the affirmative. Einstein's definition of simultaneity is a *relative* one, that is, dependent on the state of motion of the inertial system in which all the clocks rest. This dependence is then inherited by all derived notions, like, e.g. that of a length (as discussed above). This will be discussed in some detail in the next section.

Finally we wish to comment on a point of some conceptual importance. The fact that one usually adopts Einstein's definition of synchrony and simultaneity does not imply that other definitions would be somehow unphysical or inconsistent. Alternative definitions of synchrony are conceivable, which would lead to other notions of simultaneity. Here we note in passing that the already mentioned synchronization by clock-transport turns out to be equivalent to Einstein's definition in a well defined slow-transportation limit. This is explicitly demonstrated in Sect. 5.7. However, Einstein's synchrony is preferred because of a number of nice properties. In particular, it respects the principle of relativity. To see what is meant by this we note that Einstein's definition of synchrony can not only be applied to clocks at rest, but also to clocks moving at the same uniform velocity. For example, being at rest in K, one may synchronize the clocks resting in K' by the very same procedure as outlined above using light signals, now between moving clocks. But the universality of the speed of light now implies that this will be just identical to the ordinary Einstein synchronization in K'. Hence, whatever the common uniform velocity of a family of clocks is, their Einstein synchronization is always achieved by the very same prescription in terms of light signals.

This would definitely be false if instead of light other signals were used, whose propagation speed relative to an observer depends on his state of motion. This is why light is so special. The preferred rôle played by Einstein synchrony can also be characterized in a more mathematical fashion [10].

3.2 Lorentz transformations

Let K' be an inertial system that moves relative to K at constant speed v in the x direction. Both systems carry along families of identical clocks, one at each spatial coordinate position, which are synchronized according to Einstein's prescription. We ask for the transformation rules that now replace (2.2). There we identified the time parameter t with the inertial time scale reading of a clock and implicitly assumed that this reading could be instantaneously communicated to all points in space. This is how in (2.2) we (almost trivially) arrived at $t = t'$. Respecting the physical facts, we now wish to interpret t and t' according to Einstein's definition of simultaneity.

We first consider a rod (e.g. a yardstick) resting in K' as observed from K. Here and in the rest of this book it is convenient to plot ct (rather than just t) on the vertical axis, where c is the vacuum speed of light. This endows the unit on the time axis with the physical dimension of a length, like that on the spatial axes. We further agree that these units should be the same, so that world lines of light rays are depicted at an inclination of 45 degrees. In **Fig. 3.2** the world lines of the leading and trailing end of the rod are depicted by solid lines, whereas the world line m of its midpoint is dotted. Point A on the world line of the trailing end is taken as the origin ($x = 0, t = 0$) of K. ℓ denotes the world line of a light signal that at time $t = 0$ is sent off from $x = 0$ along the positive x-axis. It intersects m at M. Point E on the world line of the rod's leading end is uniquely determined by the condition that a light signal that is sent off at E in the negative x direction also intersects m at M. This, according to Einstein's definition, is just the condition for the events A and E

Lorentz transformations 47

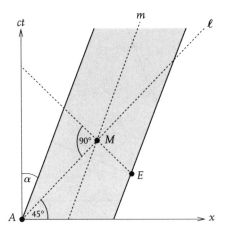

Fig. 3.2 Moving rod.

to be simultaneous with respect to time t' in K'. The construction of E suffices to determine the x'-axis, which by its very definition consists of all points simultaneous (in K') to the event A at the origin (suppressing, as usual, the y- and z-axes). But being a straight line, the x'-axis is determined by two points. The location of the ct'-axis is obvious anyway, since it is just given by the world line of the spatial coordinate $x' = 0$. In our picture it coincides with the world line of the rod's trailing end and has a slope against the ct-axis of $\tan \alpha = v/c$.

We now show that the angle between the x'-axis and the x-axis is also given by α. For this we regard the top picture of **Fig. 3.3**, in which we now represent the x'- and ct'-axes. We prolong the segment $\overline{EM}$ and denote its intersection points with the four axes by C, B, E, D. Being the prolongation of a world line of a light signal, it intersects the x- and ct-axes at 45 degrees, as indicated at C and D. The two world lines of the rod's ends are clearly parallel, so that $\overline{ME}$ and $\overline{MB}$ are equal in length; hence the pair $\overline{BC}$ and $\overline{DE}$ is also equal in length. This implies that the two shaded triangles ABC and ADE on the bottom picture of **Fig. 3.3** are congruent—they can be transformed into each other by a reflection along ℓ. So ℓ

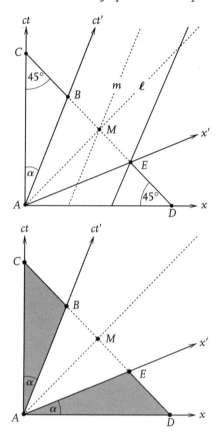

Fig. 3.3 Deriving the Lorentz transformations.

is also bisecting the angle between the x'- and ct'-axes of K'. Now we impose the condition that the speed of light measured in K' is also given by c. It immediately implies that the physical unit-length must be represented by the same geometric interval-length along the x'- and ct'-axes. Note that this does not mean that the physical unit-length is represented on the x', ct'-system of axes by the same geometric interval-lengths as for the x, ct-system. We will soon see that they must, in fact, be chosen differently.

Lorentz transformations 49

From here it is a straightforward matter to contemplate the algebraic expressions that will replace (2.2). First we recall that the transformation formulae must be linear in the coordinates, for otherwise they would not transform inertial motions—given by straight lines—to inertial motions. Moreover, in K, the x'-axis is characterized by $ct = x \tan \alpha$ and the ct'-axis by $ct = x/\tan \alpha$, where $\tan \alpha = \beta = v/c$. The first of these equations must result from the transformation formulae by setting $t' = 0$, the second by setting $x' = 0$. Hence the sought-for analytical expressions must look like

$$x' = \gamma(x - \beta\, ct) \quad \text{and} \quad ct' = \gamma(ct - \beta\, x). \tag{3.2}$$

Here γ is some yet undetermined factor which may—and will—depend on v, though only on its modulus, since otherwise a direction would be preferred. This factor precisely regulates the ratios of interval-lengths that represent the physical unit-length. It has to be the same factor in both equations of (3.2), since these interval-lengths are the same on the time and spatial axes, as we have just shown. So any transformation corresponding to a velocity v must be of that form. In particular this applies to the transformation back from K' to K, where we must replace v by $-v$. (It seems intuitively clear, and can also be proven rigorously from first principles, that if K' moves relative to K at velocity v along the x-axis, then K moves relative to K' at velocity $-v$ along the x'-axis.) Solving (3.2) for x and ct and using this condition shows that γ, as function of v, is just given by expression (2.14). Moreover, along the same lines one may show that the y and z coordinates, which we have neglected so far, transform trivially, i.e. as in (2.2). This is because they could at most be scaled according to $y' = \kappa y$ and $z' = \kappa z$, where again κ may only depend on the modulus of v. Again the inverse transformation must be of the same form so that we get $\kappa^2 = 1$ and hence $\kappa = 1$ ($\kappa = -1$ corresponds to a rotation by 180 degrees around the x-axis, which we do not consider at this point). To sum up, this yields the so-called *Lorentz transformations*, which in SR replace the old Galilei transformations:

$$x' = \frac{x - vt}{\sqrt{1 - \frac{v^2}{c^2}}}, \quad y' = y, \quad z' = z, \quad t' = \frac{t - \frac{v}{c^2}x}{\sqrt{1 - \frac{v^2}{c^2}}}. \quad (3.3)$$

For small values of v compared to the speed of light, i.e. ratios v/c small compared to one, these approximate (2.2). For larger values of v the deviations from (2.2) are, however, significant. In particular one should note that v in the Lorentz transformations is always less than c, for otherwise the analytical expressions in (3.3) become meaningless, due to the expression under the square-root in the denominator turning zero or negative. For the Galilei transformations such a restriction of the range of v would not be consistent, since the law for the addition of velocities (2.5) allows one to produce any arbitrary high velocity from the addition of sufficiently many small ones. In contrast, the restriction does make sense for Lorentz transformations, since they also imply a self-consistent modification of the addition law. This will be discussed below.

In the K' system, the coordinate values are such that, by definition, the line segment connecting the origin with the point $(x' = 1, t' = 0)$ represents a physical unit-length. According to (3.2) the latter point has coordinates $(x = \gamma, ct = \beta\gamma)$ with respect to K. These satisfy $x^2 - (ct)^2 = 1$, which describes a hyperbola which in **Fig. 3.4** is represented by the upper curve. The physical unit-length on the x'-axis is therefore given by the point where the hyperbola intersects the x'-axis. In full analogy, the physical unit of time—here measured in length—on the ct'-axis is given by the point where the positive ct'-axis intersects the hyperbola $c^2t^2 - x^2 = 1$, which in **Fig. 3.4** is represented by the lower curve. Note that according to standard Euclidean geometry the geometric lengths of the physical unit-intervals on the x'- and ct'-axes are given by $x^2 + (ct)^2 = \gamma^2(1 + \beta^2)$ (Pythagorean theorem), which is certainly larger than one. This is why these intervals look longer in our space-time diagrams, like e.g. in **Fig. 3.4**, since we intuitively always apply Euclidean distance measures to our drawings on paper. But these have no direct physical significance. The physical times measured by ideal clocks and rods obey the rules given above.

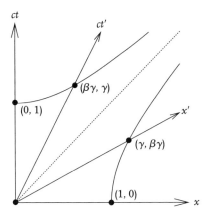

Fig. 3.4 Unit spatial lengths and times in K and K' for $\beta = 0.5$. The coordinates refer to the (x, ct)-system.

This does not mean that we may not use Euclidean geometry for our geometric discussions of space-time diagrams. It merely means that when it comes to reading off physical times and lengths from these diagrams, we may not naively identify them with Euclidean distances. Instead we have to use the rules given above, according to which unit lengths lie on hyperbolas rather than circles. One may, in fact, endow space-time with another 'geometry' in which the physical lengths and times are more directly represented. This is further discussed in Sect. 5.10.

3.3 Time dilation and length contraction

3.3.1 Time dilation

Consider a clock at rest in K', say at $x' = 0$, as seen from K. Relative to K it moves at velocity v in the x direction. At time $t = 0$ its location and reading is $x = 0$ and $t' = 0$, respectively. At time t its location is vt and its reading follows from the last equation in (3.3):

$$t' = t \cdot \sqrt{1 - \frac{v^2}{c^2}}. \tag{3.4}$$

Compared to the time t, that is defined by clocks at rest in K, the reading of the moving clock lags behind by a factor of $1/\gamma$. This is called time dilation. It is sometimes simply expressed by saying 'moving clocks slow down'. But this phrase is potentially misleading and should not be taken too literally. In that respect we stress again that all clocks involved are assumed to be physically identical. If one carefully slowed down the translational motion of the moving clock and put it next to any of the resting clocks they would run at the same rate. Evidently, due to the relativity principle, the very same statements must hold if we look at the whole situation from K'. Now the clock located at $x = 0$ moves relative to K' at velocity $-v$ along the x'-axis. The reading of this clock also lags behind by a factor of $1/\gamma$ compared to the time t' that is defined by the clocks resting in K'. This is called the reciprocity of time dilation.

The situation is depicted in Fig. 3.5, where C and C' represent world lines of clocks at rest in K and K' respectively. The clocks are adjusted to show the same time at their meeting event O. Events A on C and A' on C' are such that the same time has elapsed on both clocks since O. In K events A and B' are simultanous so that A' is reckoned later than A. On the other hand, in K', events A' and B are simultaneous so that A is reckoned later than A'.

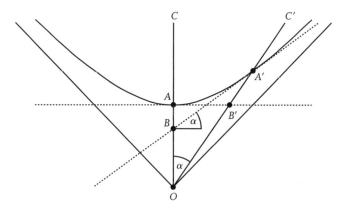

Fig. 3.5 Reciprocity of time dilation.

At first sight the statement of reciprocity may appear contradictory, like saying that if we take a clock at rest in K and a clock at rest in K', *each* runs slower than the other one. This would clearly be nonsense and is *not* what is said here. Note that in the first case we compared the reading of a single clock at rest in K' with the readings of many—at least two—clocks resting in K. Conversely, in the second case, we compared the reading of a single clock at rest in K with the readings of many—at least two—clocks resting in K'. Hence the sets of clocks involved in these two cases differ.

To put this important point even more explicitly, we denote by C_1 and C_2 two clocks at rest in K, and by C'_1 and C'_2 two clocks at rest in K'. Their world lines are depicted in **Fig. 3.6**. The events where the various world lines cross are denoted by A, B, D, and E. For reasons of pictorial simplicity we have arranged the relation between the distances of the clocks in their respective rest systems in such

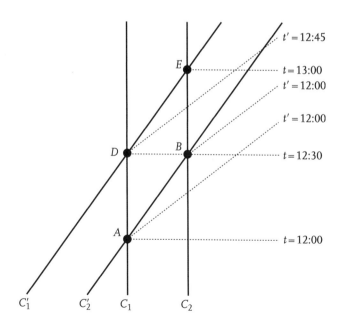

Fig. 3.6 Reciprocity of time dilation involves more than two clocks.

a way that the event where C_1 meets C_1' is simultaneous in K to the event where C_2 meets C_2'. (As will become clear after the discussion of length contraction, this means that the distance of both couples are the same when measured in K. In other words, the distance between C_1' and C_2' measured in K' is longer by a factor of γ than the distance between C_1 and C_2 measured in K.) The picture corresponds to a value of $\gamma = 3/2$, i.e. v/c approximately equals 0.75.

An observer resting in K compares the readings of both clocks, C_1 and C_2, with those of a single clock resting in K', say C_2'. At the event A the clock C_2' meets C_1 and both are adjusted to read 12:00. Then C_2' meets C_2 at the event B and C_2 reads 12:30. This last reading is also arbitrarily set by choosing the common rate of all clocks appropriately. This being done, all other readings are determined. The reading of C_2' at B is 12:20, corresponding to a factor of $1/\gamma = 2/3$ by which the moving clock lags behind. Now we take the position of an observer resting in K'. He compares his clocks C_1' and C_2' with a single clock at rest in K, say C_1. Event A again denotes the meet of C_1 and C_2' where both clocks read 12:00. At the event D clock C_1 reads 12:30 whereas clock C_1' reads 12:45, again corresponding to a factor $1/\gamma = 2/3$ by which the moving clock, which is now C_1, lags behind. Note that the first statement, asserting the 'lagging behind' of C_2' as compared to the time t in K, involves the clocks C_1, C_2, and C_2', whereas the second statement, asserting the 'lagging behind' of C_1 as compared to time t' in K', involves the clocks C_1', C_2', and C_1. This we summarize in Table 3.1.

3.3.2 Length contraction

Similar to the statements just made for measurements of time intervals are those that apply to lengths. Again we consider a rod

Table 3.1

Event	Happening	Time t of K	Time t' of K'
A	clock C_1 meets clock C_2'	12:00	12:00
B	clock C_2 meets clock C_2'	12:30	12:20
D	clock C_1 meets clock C_1'	12:30	12:45
E	clock C_2 meets clock C_1'	13:00	13:05

Time dilation and length contraction

in K' whose end points rest at $x' = 0$ and $x' = l'$. Its length in K' is defined with respect to the yardsticks that also rest in K' and is therefore just given by l'. With respect to K the rod moves at velocity v along the x-axis. Its 'length' with respect to K is defined to be the spatial distance between two simultaneous positions of its end points. Here 'distance' is understood to be measured with yardsticks resting in K and 'simultaneous' is meant with respect to the time t defined in K. For example, one may choose $t = 0$ so that the position of the rod's left end is $x = 0$. The simultaneous position $x = l$ of the right end is obtained from the first equation of (3.3) by setting $x' = l'$ and $t = 0$:

$$l = l' \cdot \sqrt{1 - \frac{v^2}{c^2}}. \tag{3.5}$$

Hence with respect to the length measure defined in K the rod is shorter by a factor of $1/\gamma$, as compared to its length in K'. Note that again we may assume all yardsticks and rods to be physically of the same constitution. If one slowed down the rod and placed it next to an equivalent one that rests in K there would be no difference in length. By the principle of relativity it is also true that a rod resting in K appears shorter by the same factor when measured from K'. Again this may sound contradictory at first, since it seems to claim that of two physically equivalent rods resting in K and K', respectively, *each* is shorter than the other one. This would indeed be nonsensical if 'being shorter' referred both times to the very same notion of length. But again this is *not* the case in the situation at hand. In the first situation, length refers to the distance between simultaneous positions of the rod's end points with respect to simultaneity in K (t-time), in the second case with respect to simultaneity in K' (t'-time). These two notions of length differ in the operational sense. We will illustrate this point shortly by means of space-time diagrams.

The state of affairs expressed in (3.5) is called length contraction (alternatively Lorentz or Lorentz–FitzGerald contraction). It is often summarized by saying that 'moving bodies shrink in length' (in the direction of motion). Again, as follows from the reciprocity

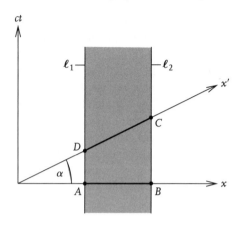

Fig. 3.7 World surface of a rod resting in K.

of length contraction, this should be properly understood in order to avoid confusion.

Let us now interpret the length contraction by means of the space-time diagram in **Fig. 3.7**. Let there be a (idealized one-dimensional) rod of length l resting in K. The world lines of its two ends are ℓ_1 and ℓ_2. The shaded surface consists of all events at which the rod 'exists', meaning that at this time and spatial location there is some matter element of the rod. This set of events is called the rod's 'world surface'. Those points of this surface which are simultaneous in K at $t = 0$ form the interval $\overline{AB}$. One may call this one-dimensional set of points the 'rod at time $t = 0$'. By an equivalent argument the interval $\overline{DC}$ may, with the same right, be called 'the rod at time $t' = 0$'. Note that although $\overline{DC}$ appears longer than $\overline{AB}$ by a factor of $1/\cos\alpha = \sqrt{1 + \beta^2}$ in our picture, it corresponds, in fact, to a physical length shorter by a factor of $1/\gamma$. As discussed above, this is because on the x'-axis the intervals corresponding to the physical unit-length are longer by a factor of $\gamma\sqrt{1 + \beta^2}$ as compared to such intervals on the x-axis. A space-time viewpoint of the world therefore suggests that the length contraction is a mere effect of projection (from four-dimensional space-time into three-dimensional space), similar to those effects of perspective that are familiar from

Time dilation and length contraction 57

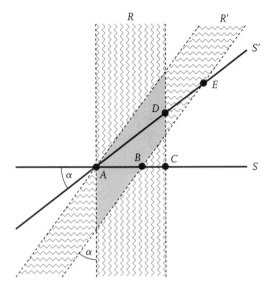

Fig. 3.8 Reciprocity of length contraction.

two-dimensional pictures of three-dimensional objects. There the three-dimensional geometry is absolute, given the body, but the two-dimensional projections vary according to the line of sight. Here, in SR, the four-dimensional world is absolute, but its splitting into space and time and therefore the three-dimensional projections into 'things at a time' are relative, depending on the state of motion of the observer.

Finally, as promised, we interpret the reciprocity of the length contraction by means of a simple space-time diagram. In **Fig. 3.8** the world surfaces of two rods, R and R', are shown as they present themselves in the rest system K of R. The rods are chosen to be of the same length in their respective rest systems, so that in our picture the horizontal section across R' is shorter by a factor of $1/\gamma$ than the horizontal section across R. In the picture γ is again chosen to be 3/2, corresponding to a velocity approximately 75% of the velocity of light; hence $\tan \alpha \approx 0.75$ or $\alpha \approx 37°$. The lines S and S' correspond to simultaneous events in K (constant t-time) and K'

(constant t'-time), respectively. Hence the events A, B, and C are simultaneous in K whereas A, D, and E are simultaneous in K'. The intervals $\overline{AC}$ and $\overline{AB}$ correspond to the lengths of R and R', respectively, as measured in K. Obviously $\overline{AB}$ is shorter than $\overline{AC}$. On the other hand, the intervals $\overline{AD}$ and $\overline{AE}$ correspond to the lengths of R and R', respectively, as measured in K'. Now $\overline{AD}$ is obviously shorter than $\overline{AE}$. One may easily prove that the ratios of the shorter to the longer interval length is the same in both cases, namely $1/\gamma$.

3.4 Addition of velocities

In this section we wish to discuss the modification that the old law (2.5) for the addition of velocities receives in SR. We have already seen that the old law would be inconsistent with the transformation formulae (3.3), which become mathematically singular for speeds greater than or equal to c. To deduce the new law, we think of a projectile which moves relative to K' with a velocity $\vec{u}' = (u'_x, u'_y, u'_z)$. The analytical expression for its motion is again given by (2.3). Introducing this into (3.3) one obtains expressions of the form $x = u_x t$, $y = u_y t$, and $z = u_z t$, where

$$u_x = \frac{u'_x + v}{1 + u'_x v/c^2}, \quad u_y = u'_y \cdot \frac{\sqrt{1 - v^2/c^2}}{1 + u'_x v/c^2}, \quad u_z = u'_z \cdot \frac{\sqrt{1 - v^2/c^2}}{1 + u'_x v/c^2}.$$
(3.6)

This is a surprisingly complicated law, totally different from the usual addition of vectors, except that it approximates vector addition in the limit of small v/c. Indeed, this new addition law defines an operation which constructs three new velocity components $\vec{u}$ relative to K out of the three components for the velocity $\vec{u}'$ relative to K' and the three components for the velocity $\vec{v}$ of K' relative to K. We denote this operation by $\oplus$, i.e. we write $\vec{u} = \vec{v} \oplus \vec{u}'$. It is not hard to see that this operation is neither commutative nor—what is worse—associative. That means that we generally neither have $\vec{v}_1 \oplus \vec{v}_2 = \vec{v}_2 \oplus \vec{v}_1$ nor $\vec{v}_1 \oplus (\vec{v}_2 \oplus \vec{v}_3) = (\vec{v}_1 \oplus \vec{v}_2) \oplus \vec{v}_3$, as we are used to from ordinary addition.

It is quite easy to deduce from (3.6) that the modulus of $\vec{u}$ is always smaller than c, given that this is true for $\vec{u}'$ (it holds for $\vec{v}$ in any case). Hence it is impossible to get equal to, or even above, the velocity of light by successively adding velocities smaller in modulus than c. For example, adding together half the velocities of light in the x direction according to the first formula in (3.6) does not result in c but $4/5c$, as one easily verifies. Now, every inertial system can be obtained from a given one by some application of a Lorentz transformation of the form (3.3), a translation, and a spatial rotation. But the latter two have no effect on the modulus of velocities. This implies that the modulus of the velocity of a projectile is less than c in any inertial system if this holds in at least one. In particular, a 'rest system' can always be found for such a projectile.

Similar statements hold for moduli greater than c, since (3.6) also implies that the modulus of $\vec{u}$ exceeds c if that of $\vec{u}'$ does (the modulus of $\vec{v}$ is, as always, still less than c). Here it is even possible to formally produce infinite velocities. For example, consider a fictitious process propagating in K' with superluminal speed $u'_x > c$ in the x direction. Its velocity relative to K will turn out to be infinite, according to (3.6), if K is chosen such that $v = -c^2/u'_x$. Note that the minus sign says that K' moves relative to K in the direction of the negative x-axis, so that K *follows* the signal. This says that running after a superluminally propagating projectile enhances rather than diminishes its relative speed, until it becomes infinite at a critical (subluminal) running speed at which the denominators of (3.6) become zero. Running still faster, though always less than c, makes the relative projectile's velocity change sign, since now the the denominators in (3.6) turn negative. The projectile now approaches the observer at superluminal velocities! All that sounds of course pretty fantastic and merely indicates that, within SR, all processes that rely on causal relationships cannot propagate at superluminal velocities. These, in particular, comprise those processes which can be used for the transmission of information and/or energy, that is, all processes which one may use for the propagation of signals. But note that this does not imply the absence of *any* kind of superluminal velocities. For more discussion on this point see Sect. 5.3.

Finally we present a simple but convincing application of the addition law for velocities to Fizeau's experiment that we discussed in Sect. 2.4.2. Let K' be the rest system of the medium within the tube, relative to which light travels along the x'-axis at speed $u'_x = c/n$. Let, further, K be the laboratory system relative to which the medium flows at speed v along the x-axis. Then the first equation in (3.6) predicts the speed u_x at which the light propagates in the laboratory system K:

$$u_x = \frac{v + c/n}{1 + v/(cn)} \approx \frac{c}{n} + v\left(1 - n^{-2}\right). \tag{3.7}$$

Here the $\approx$ symbol indicates an approximation, which consists of neglecting terms that are suppressed relative to those written down by a factor of v/c, or higher powers thereof. In Fizeau's and related experiments v was smaller than 10 metres per second, which means that v/c is smaller than $3 \cdot 10^{-8}$. Hence the above approximation is fully justified. Now note that (3.7) is just identical to (2.10) and (2.11). The somewhat mysterious 'dragging coefficient' thus turn out to be a simple consequence of the new addition law, and not an expression of a complicated interaction between matter and some hypothetical ether.

3.5 Causality relations

We have seen that within the kinematical framework of SR no causal dependencies may propagate faster than by the speed of light. Here we wish to give a geometric interpretation of this important consequence by means of space-time diagrams. Such a representation was not included in Einstein's original paper of 1905. It goes back to the mathematician Hermann Minkowski (1864–1909), who in his famous address 'Space and Time' of 1908 [11] first pointed out the usefulness of this representation technique. In his honour space-time diagrams are therefore sometimes also called Minkowski diagrams. Minkowski also made other seminal contributions to SR that led to its modern powerful

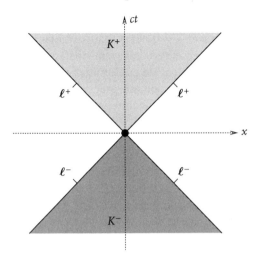

Fig. 3.9 Light cone and domains of causal dependence of the event •.

mathematical formulation, which, however, we shall not make use of in this book.

Let K be an inertial system whose (ct, x)-axes are depicted in **Fig. 3.9**. Again two dimensions (the y- and the z-axis) are suppressed. Hence the wave front of a flash of light that sparked off at $x = 0$, $t = 0$ is represented by the two world lines denoted by ℓ^+, one each for the propagation along the positive and negative x-axis, respectively. The union of these two half-lines is called the future light-cone ℓ^+ of the event $O = (x = 0, ct = 0)$ (depicted by • in our picture). It consists of all points in space-time that can receive light from O. It is called a 'cone' since if one adds a further spatial dimension one obtains a three-dimensional space-time diagram which can be obtained from our two-dimensional version through rotation about the ct-axis. The surface so generated by ℓ^+ is a two-dimensional cone whose vertex lies at O; see the figure used as frontispiece on Page ii. After addition of one more space dimension one finally obtains a three-dimensional cone in four-dimensional space-time, which is clearly much harder to visualize.

The light cone contains all events that can be reached by a light signal originating from O. The inner domain of the cone, K^+, depicted in a lighter shading, consists of all events that can be reached from O by processes that propagate subluminally. This is because the coordinates (ct, x) of this domain precisely satisfy the condition $|x| < ct$. The region K^+ is called the *chronological future* of O. The union of K^+ with the future light cone is called the *causal future* of O, since it just consists of all events that O can causally influence, either by subluminal or luminal propagation.

Similarly one considers the light rays which at time $t = 0$ meet at $x = 0$. The union of these half-lines, ℓ^-, is called the past light cone of O. It consists of all points of space-time which can send light signals to O. Its interior domain, K^-, depicted in a darker shading, contains all points which can reach O by processes that propagate subluminally. It is called the *chronological past* of O. The union of K^- and the past light cone is called the *causal past* of O. It consists of all events that can causally influence O.

The third region of interest is that outside the causal future and past of O. It consists of all events that can neither be causally influenced by O nor themselves causally influence O. For this reason it is called the *causal complement* of O. The existence of such causal complements is a direct consequence of a finite upper bound for all signal velocities and a new feature that SR introduces into our space-time concepts. For events lying in the causal complement of O it makes no absolute sense to say that they happened before or after O. This is because a Lorentz transformation can make the new x'-axis any straight line through O in its causal complement. Hence for any event E in the causal complement of O there exist an inertial system K' in which E is assigned a smaller value of t' than O, but likewise there also exists an inertial system K'' in which E is assigned a larger t'' value than O. In K' event E would be said to happen after O, using inertial time t', in K'' the time order would be just opposite with respect to inertial time t''. This is illustrated in **Fig. 3.10**. Note that the situation regarding time orders is totally different in the causal future and past of O. Any inertial observer agrees that events in K^+ happen after and events in K^- before O.

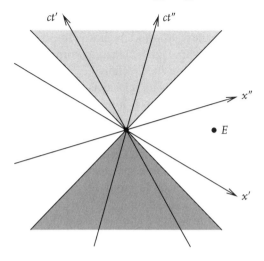

Fig. 3.10 Events E in the causal complement of ● have no Lorentz-invariant relative time order.

3.6 Aberration and Doppler effect

The addition law (3.6) also predicts that the modulus of $\vec{u}$ equals c if and only if the modulus of $\vec{u}'$ equals c. This holds independently of $\vec{v}$, whose modulus, we recall, is always less than c. This is immediate from the very definition of the Lorentz transformations, one of whose two axioms were that the velocity of light should be the same in all inertial systems. Hence only the direction and frequency of a light ray may vary from one to another inertial observer, depending on the relative velocity between observer and light source. These velocity-dependent variations of directions and frequencies are called aberration (cf. Sect. 2.4.1) and Doppler effect respectively. Their laws will now be derived.

3.6.1 Aberration

We consider a light source, L, resting in the $x'y'$-plane of the inertial system K'. An observer is situated at the origin of K' who receives and analyses the light being sent to him from L. Let α' be the angle enclosed between the x'-axis and the line connecting the observer to L (i.e. his line of sight). Light emanating from L travels along this line at speed c in the opposite direction, i.e. from L to the observer. Hence the x'-component of its velocity is given by $u'_x = -c\cos\alpha'$. Let us now assume that there is a second observer at the origin of the inertial system K, relative to which K', and hence L, moves with velocity v along the x-axis. The phenomenon of aberration now boils down to the statement that even though at time $t = t' = 0$ both observers are at the same space point, they nevertheless do not measure the light as coming from the same direction. To put this quantitatively, let α be the angle between the line of sight along which the observer in K sees L and his x-axis. The x-component of the light-signal's velocity in K is likewise given by $u_x = -c\cos\alpha$. Note that here we made use of the universality of the speed of light. Inserting these expressions for u_x and u'_x into the first equation (3.6) immediately yields an expression relating the cosines of α and α' in a v-dependent fashion (β and γ are as in (2.14)):

$$\cos\alpha = \frac{\cos\alpha' - \beta}{1 - \beta\cos\alpha'}. \tag{3.8}$$

This is already the relativistic aberration formula. Sometimes it is more convenient to rewrite it in terms of the tangents (instead of the cosines) of half the angles:

$$\tan\frac{\alpha}{2} = \tan\frac{\alpha'}{2} \cdot \sqrt{\frac{1+\beta}{1-\beta}}. \tag{3.9}$$

Since the tangent of $\alpha/2$ is a monotonic function of α within the relevant domain between 0° (source directly moving away from observer) and 180° (source directly approaching the observer), it is easy to tell straightaway the qualitative behaviour of (3.9). Imagine a spaceship which moves with increasing speed toward some fixed star S. Let S' be the star (suppose there is one) diametrically opposite to S on the celestial sphere. We take S and S' to be the poles of that sphere. Then, as the spaceship's speed increases, the astronaut will see the stars moving away from S' and toward S, along the meridians that connect the poles S' and S of his celestial sphere. The poles themselves remain fixed.

3.6.2 Doppler effect

We now turn to the Doppler effect. We assume the source L to emit monochromatic light of frequency v', as measured in the rest system K' of L. This means that the time interval measured in K' (i.e. t'-time) between two equal phases of light is given by $\tau' = 1/v'$. Due to time dilation (3.4), this corresponds to the longer interval $\tau = \gamma \tau'$ in K (i.e. t-time). During this time the distance between L and the observer at the origin of K increases by the amount $v\tau \cos \alpha$. The light phase sent off from L at the end of the interval τ must therefore travel this extra distance and hence arrives with a delay of $\beta \tau \cos \alpha$ as compared to the light phase sent off just at the beginning of the time interval. (To obtain the last expression we again used the universality of the speed of light.) This delay has to be added to τ in order to get the period of the light wave as measured by the observer in K. Finally, since the frequency is just the inverse of the period, we obtain the following expression for the frequency measured by the observer in K:

$$v = \frac{v'}{\gamma(1 + \beta \cos \alpha)}. \tag{3.10}$$

This is the special-relativistic law for the Doppler effect. Because of aberration, it is important to be aware that this formula assumes a different form if written in terms of the angle α' instead of α. The difference is a second order effect in v/c and becomes crucial for the transverse Doppler effect that we discuss below. Here we have chosen to write the law in terms of α, since this is the angle that the observer in K measures between his direction of sight and the velocity of L. In contrast, α' would be the angle that a co-moving observer at the position of the source L would measure between the emitted light ray that travels toward the origin of K and the velocity of K relative to K'. The corresponding relation is obtained by replacing $\cos\alpha$ in (3.10) by the expression (3.8). It will be stated and used later; see (3.19).

Except for the appearance of γ expression (3.10) was already part of 'pre-relativistic' physics. There it was derived just as above, even though we used the universality of the speed of light. But, in fact, the only assumption we actually used was that the velocity of light was isotropic and of modulus c in K. Hence if we identify K as the rest system of a medium (air, water) in which wave propagation takes place, our derivation above applies almost verbatim to sound or water waves. Only the initial argument concerning the factor γ is now simply omitted and c is interpreted as the propagation velocity of waves in the specific medium under consideration. Formula (3.10) without γ therefore also gives the correct expression for sound or water waves, whenever the source moves relative to the medium and the observer is at rest. In the opposite case, where the source is at rest and the observer moves with respect to the medium, one would have obtained a slightly modified expression in which the factor $(1+\beta\cos\alpha)$ in the denominator of (3.10) would be replaced by a factor $(1-\beta\cos\alpha)$ in the numerator. Hence the Doppler effect for waves in media depends not only on the relative velocity between source and observer, but also on their absolute velocity relative to the medium. This would therefore also be the case in an ether theory of light without time dilation. It is the essence of SR to have overcome this distinction. Accordingly, equation (3.10) only depends on the *relative* state of motion between observer and source. There is no reference to a hypothetical ether system which would be distinguished by being the only system in which the velocity light did not depend on the direction.

An important special case of (3.10) is given for the case where the observational direction is perpendicular to the velocity of the source, i.e. where $\alpha = 90°$, so that $\cos\alpha = 0$. Then (3.10) reduces to

$$\nu = \nu' \cdot \sqrt{1 - \frac{v^2}{c^2}}. \qquad (3.11)$$

Here the difference between ν and ν' is solely given by the γ-factor, whose existence at this point is exclusively due to time dilation. The frequency shift at perpendicular observational directions is called the 'transverse Doppler effect'. Already in 1907 Einstein suggested experimentally verifying the appearance of γ in (3.10), and hence the existence of time dilation, through direct observation of the transverse Doppler effect [12]. But this was only achieved much later (1938) in the experiment of Ives and Stilwell, which we will discuss in Sect. 5.5. Finally we come back to our previous remark concerning aberration and the different rôles that α and α' play in the law for the Doppler effect. If instead of α we had set α' equal to 90°, which according to (3.8) is equivalent to setting $\cos\alpha = -\beta$, we would have obtained the inverse factor on the right-hand side of (3.11). Hence observing a moving light source perpendicular to its direction of motion results in a 'red-shifted' (i.e. lower) frequency by a factor of $1/\gamma$, whereas the light that the source emits perpendicular to the direction of motion of the observer reaches him at a 'blue-shifted' (i.e. higher) frequency by a factor of γ. Only the first case is usually referred to as the 'transverse' Doppler effect.

3.7 Length contraction and visual appearance

The visual appearance of an object is given by the light signals that *simultaneously* enter the eye of the observer or the lens of his photographic device. But the different parts of an extended body vary in distance to the observer's eye. Hence the parts further away must send their light earlier than the closer lying parts. What an inertial observer sees are therefore not the mutually simultaneous (with respect to his inertial time) parts of the object. On the other hand, the geometry of a body is defined by the simultaneous positions of its parts. This is how we defined the 'length' of a body for which we derived length contraction. What, then, does a moving body look like? Can we see the length contraction at all?

In general, the determination of the apparent shape of a body is a very complicated analytical problem. But we can simplify matters

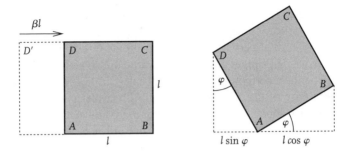

Fig. 3.11 Visual impression of a moving cube. The observer is located in the paper plane far below the cube.

considerably by restricting attention to a body whose extent is small in comparison to its distance to the observer. In this case we may, in a first approximation, treat all light rays from the body to the observer's eye as parallel. Furthermore we assume that at the moment of observation the direction of sight from the observer to the object is perpendicular to the velocity of the latter. This corresponds to $\alpha = 90°$ (like in the transverse Doppler effect) so that (3.8) implies $\cos \alpha' = \beta$. But α' is just the angle between the light rays emitted by the object and the direction of the moving observer, i.e. the negative x'-axis. Hence the object does not show its 90° sideview to the observer but rather a slightly rotated one. The rotation angle is $\varphi = 90° - \alpha'$, which satisfies $\sin \varphi = \beta$ and $\cos \varphi = 1/\gamma$.

Besides a rotation, the object may also appear deformed. To clarify this in a simple example we further assume the object to be a cube, the top-view of which we depict in **Fig. 3.11**. The observer, who is not represented in the picture, is located in the plane of the paper a good distance below the cube. The cube's edge length at rest is l. Light emitted from its far left corner, D, needs to travel a distance l farther than light from A before both reach the observer. Hence it must be emitted a time interval l/c earlier in order to reach the observer's eye at the same time as the light emitted from A. At this earlier time the cube was located a distance $vl/c = l\beta$ further to the left. Hence the observer can actually see the side-edge

$\overline{AD}$ as projection of length $l\beta$ perpendicular to his line of sight. Its appearance is as if the cube were rotated by an angle φ, where $\sin\varphi = \beta$. Because of length contraction, the observer sees the edge $\overline{AB}$ (all points of which are taken to have the same distance from the observer in our approximation) with length $l/\gamma = l\cos\varphi$, again as if the cube where rotated by the angle φ. This means that the observer sees a *rotated but undeformed* cube. If there were no length contraction, so that the observer were to see $\overline{AB}$ with length l, he would have the impression of a rectangular solid of length γl, height and depth l, seen rotated by an angle φ. The somewhat ironic punchline of this is that *because* of length contraction we see the cube rotated and undeformed rather than rotated and prolonged.

The general laws for visual images are far more involved, in particular if the object dimensions are not small compared to the object's distance. But one may still rigorously prove that the image of a moving ball is still a ball and not, as one might naively think, an oblate spheroid. Moving balls or spheres do not look contracted in their direction of motion. The general proof of this is explained in Sect. 5.8. It seems strange that the distinction between the visual appearance of a body on one side, and the geometric shape as defined by the simultaneous positions of the parts of the body on the other, was not recognized for a long time. The first seems to have been Anton Lampa in 1924 [13], but his paper apparently went unnoticed until recently. In 1959 essentially the same observations were made again in greater detail by James Terrell [14] and, in the special case of the moving sphere, by Roger Penrose [15]. Instructive and amusing animations can be found on the INTERNET, e.g. [16, 17].

3.8 Mass, momentum, and kinetic energy

Mechanics is conceptually deeply linked with our notions of space and time. Any modifications of the latter, as, e.g. brought about by SR, will necessarily also affect mechanics. This will be explained in some detail in this section.

A central theorem in Newtonian mechanics states the conservation of momentum. In the special case of two colliding bodies it takes the following form. Let $\vec{u}_1$ and $\vec{u}_2$ be the velocities of two bodies before collision. After an elastic collision the velocities have changed and are now given by $\vec{v}_1$ and $\vec{v}_2$, respectively. The bodies may mutually exchange any kind of forces, but we assume that no external forces act on the system. The theorem then states that there are two numbers (together with a physical dimension, like 'kilogram'), m_1 and m_2, so that

$$m_1\vec{u}_1 + m_2\vec{u}_2 = m_1\vec{v}_1 + m_2\vec{v}_2. \qquad (3.12)$$

The quantities m_1 and m_2 are associated with the individual bodies and are independent of their states of motion. Hence (3.12) is a universal law that for given initial velocities $\vec{u}_1$ and $\vec{u}_2$ restrict the possible outcomes $\vec{v}_1$ and $\vec{v}_2$. The quantities m_1 and m_2 are the (inertial) masses of the bodies which govern their inertial behaviour, like, e.g. the centrifugal force. The product of mass times velocity is called the 'momentum' of the body. Equation (3.12) states that the sum of all momenta before collision equals the sum of all momenta after collision, i.e. that momentum is conserved. We also recall that Newtonian force is, strictly speaking, defined as the time rate of change of momentum, i.e. its time derivative. Replacing momentum by the product of mass times velocity, and assuming the mass to be constant in time, leads back to the well known equation (2.1).

We now ask: what in SR is the expression for momentum as a function of mass and velocity? We do not yet know the answer, but we can uniquely deduce it through the requirement of momentum conservation. That is, we *define* momentum so as to be conserved in time. Moreover, for small velocities the new expression for momentum should approximate the Newtonian one, given by the product of mass times velocity. On the other hand, it is also easy to show that (3.12) cannot be strictly valid in SR. Hence the Newtonian expression for momentum truly has to be modified in order to save the law of momentum conservation into SR. An obvious strategy is to assume the validity of (3.12) under the relaxation that m_1 and

Mass, momentum, and kinetic energy

m_2 may also depend on the respective velocities. This dependence should just involve the moduli of the velocities in order not to prefer any direction in space. We therefore assume the momentum $\vec{p}$ of a body, moving with velocity $\vec{u}$, to be given by an expression of the form $m(u)\vec{u}$, where $m(u)$ is a yet unknown function of the modulus u of $\vec{u}$. We now show that this function is uniquely determined by the requirement of momentum conservation. For the time being, and merely for convenience, we follow some old fashioned terminology and continue to call m the 'mass', even though it is no longer a fixed number. We will explain the modern terminology below.

Again we introduce two inertial systems, K and K', where K' moves relative to K at speed v in the x direction. At $t = t' = 0$ both coordinate systems coincide. We consider the elastic scattering of two physically identical spherical bodies A and B. Let this scattering be arranged as follows. Before the scattering event, A moves relative to K at speed w on the negative y-axis in a positive (upward) direction, whereas B moves relative to K' at the same speed on the positive y'-axis in a negative (downward) direction. Note that both speeds refer to different inertial systems. The timing of these motions is such that at time $t = t' = 0$ the bodies collide at the (then coinciding) origins of K and K'. At this moment (and only then) the bodies mutually exchange forces whose direction must be along the line connecting the centres of the bodies at this moment, i.e. along the (then coinciding) y- and y'-axes. We assume the collision to be totally elastic, which means that no deformations of the bodies take place and hence no energy is lost. This implies that after collision the moduli of the velocities are the same as before with exactly inverted direction. Hence, after collision, A moves relative to K on the negative y-axis at speed w in a negative (downward) direction and B moves relative to K' on the y'-axis with the same speed in the positive (upward) direction.

Let us now describe the whole process relative to system K. Here A has the velocities as stated above. To obtain the velocities of B relative to K we have to use the rules (3.6). Before collision we get $u_x = v$ and $u_y = -w/\gamma$; after collision $u_x = v$ and $u_y = w/\gamma$. This

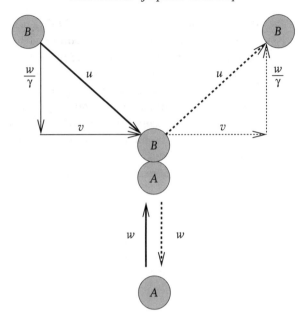

Fig. 3.12 Collision of A and B as reckoned from system K.

is summarized in **Fig. 3.12**, where the velocities before collision are marked by solid arrows, after collision by dashed arrows. The thin arrows signify the just cited horizontal and vertical components of B's velocities, which just add to the velocity vectors as indicated. The modulus of the velocity vectors is given by

$$u = \sqrt{v^2 + (w/\gamma)^2}. \tag{3.13}$$

Since the mass is assumed to be a function of the modulus of the velocity only, it follows that the masses of A and B do not change during the collision process. For A it is given by $m(w)$ and for B by $m(u)$, where u stands for the expression in (3.13). The x-component of total momentum is just that due to B, which is obviously conserved. For A as well as for B we have that the y-component after collision is just the negative of its value before collision. It follows that this is also true for their sum. But the sum should be preserved, i.e. be the same before and after collision. Hence the value of the sum must be zero: $m(u)w/\gamma - m(w)w = 0$. Momentum conservation in the present example is therefore equivalent to

$$m(u) = \gamma m(w), \qquad (3.14)$$

where u stands again for the expression in (3.13). This equation must hold true for all values of v and w (recall the v-dependence of γ). Hence, in particular, it must hold for arbitrary small values of w and, by continuity, also for $w = 0$. In this limiting case u equals v and we get (writing out γ explicitly)

$$m(v) = m_0 \gamma = \frac{m_0}{\sqrt{1 - v^2/c^2}}, \qquad (3.15)$$

where we set $m_0 = m(0)$. This is a unique prediction of m as a function of speed. We obtained it from (3.14) by going through a limiting case. But one may now straightforwardly verify that it indeed solves (3.14) in full generality.

We are now ready to write down the expression for momentum in SR:

$$\vec{p} = m(v)\vec{v} = m_0 \gamma \vec{v} = m_0 \frac{\vec{v}}{\sqrt{1 - v^2/c^2}}. \qquad (3.16)$$

The quantity m_0 denotes the mass at zero speed. It is therefore called the 'rest mass'. It is this quantity which most closely characterizes the 'amount of substance'. In that respect it resembles the Newtonian mass more closely than our mass function (3.15), though in the next section we will see that m_0 does vary with the internal state of the body. A Newtonian reading of (3.16) suggests calling that quantity 'mass' that multiplies $\vec{v}$, i.e. $m(v)$. This used to be standard terminology in older textbooks. It is suggestive for Newtonian intuition, in that it explains the impossibility to accelerate a body to or beyond the speed of light by its unboundedly increasing mass. In daily life this increase of mass is absolutely negligible. For example, a speed of 200 kilometres per hour would result in a fractional mass increase of 10^{-14}. A doubly-supersonic flight pushes this up by two orders of magnitude. But this still makes a minute effect of one in a billion (British) or trillion (American); in any case, one in 10^{12}.

The situation is entirely different in elementary particle physics (cf. Sect. 4.3). Here, in the extreme case, γ-factors of more than a thousand are reached, which means that the mass also increases by that factor. This enormous variability makes it clear that this notion of mass is not suitable to intrinsically characterize elementary particles. Here, in elementary particle physics, it is therefore much more natural to identify the 'mass' of a particle with its rest mass. This now corresponds to the generally adopted way of speaking.

It is a routine calculation, using elementary calculus, to determine the kinetic energy as a function of velocity, once momentum as a function of velocity is determined. To understand this, we recall that the force that one needs to apply in order to accelerate the body is just given by the time derivative of the body's momentum. On the other hand, the work done during the acceleration process is given by the integral over the force along the path along which the body moves. Because of energy conservation, the work done, starting from zero velocity, is numerically the same as the kinetic energy of the body after acceleration. Starting from (3.16), this leads to

$$E_{\text{kin}} = m_0 c^2 (\gamma - 1) = m_0 c^2 \left(\frac{1}{\sqrt{1 - v^2/c^2}} - 1 \right). \qquad (3.17)$$

As expected, the kinetic energy grows unboundedly if the speed approaches c. For small speeds one has $\gamma \approx 1 + v^2/2c^2$ and hence $E_{\text{kin}} \approx \frac{1}{2} m_0 v^2$, which is just the Newtonian expression. **Figure 3.13** compares the plots for the Newtonian (solid curve) and special relativistic (dashed curve) expressions of kinetic energy, here measured in units of the rest energy $m_0 c^2$.

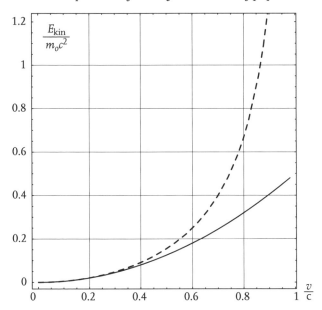

Fig. 3.13 Kinetic energy as a function of speed according to Newtonian mechanics (solid curve) and special relativistic mechanics (dashed curve).

3.9 Probably the most famous formula in all of physics

The next volume of *Annalen der Physik* to that in which Einstein published his work on special relativity contains a small supplement to this paper, hardly three pages long, entitled 'Does the inertia of a body depend on its energy content?' [18]. Here 'energy content' refers to purely internal energy, not kinetic energy. That kinetic energy contributes to inertia is already clear from the velocity dependence of the mass (3.15). This can be made more explicit by writing the mass as a function of the kinetic energy:

$$m = m_0 + \frac{E_{\text{kin}}}{c^2}. \qquad (3.18)$$

The following argument of Einstein's aims to also connect the rest mass m_0 with the energy content of the body. We consider a body A that rests in the inertial system K'. At time $t' = 0$ it emits energy in the form of light. To simplify the ensuing discussion we will use the quantum theoretic picture, according to which the energy of light is concentrated into spatially localized 'quanta', or 'photons' as they are now called. This light-quantum hypothesis was introduced by Einstein in the same year 1905, but not used by him in the present context. So let us assume that two photons of equal frequency ν' (measured in K') are emitted in diametrically opposite directions by A. According to Einstein's light-quantum hypothesis, each photon carries a momentum $h\nu'/c$ (relative to K') in the direction of propagation. Here h denotes Planck's constant. Momentum conservation then implies that the emission of such a photon causes a recoil (momentum transfer) of equal strength but opposite direction on to the emitting body. Since two photons of equal frequency are emitted in opposite directions, both recoils cancel and A does not change its state of motion relative to K', i.e. stays at rest. Even though the momentum of A relative to K' does not change, its energy does. Again the light-quantum hypothesis predicts that each photon carries an energy of $h\nu'$ (relative to K'). Energy conservation then demands that A loses the amount $\Delta E' = 2h\nu'$ of energy, as measured in K'. The upper picture of **Fig. 3.14** shows the emission process relative to K'.

Now we regard the same process in the inertial system K, relative to which K' moves with velocity v in the x direction. This corresponds to the lower picture of **Fig. 3.14**. Since A is at rest in K' before and after emission, it keeps its velocity v relative to K. The latter is true despite the fact that the two photon emissions are not directed oppositely in K. The latter is just a consequence of aberration and can be easily checked using (3.9). Moreover, the Doppler effect implies that the frequencies of the two photons differ in K. To see this quantitatively, it is more convenient in the present context to use (3.8) to re-express (3.10) in terms of α', which is the angle

Probably the most famous formula in all of physics 77

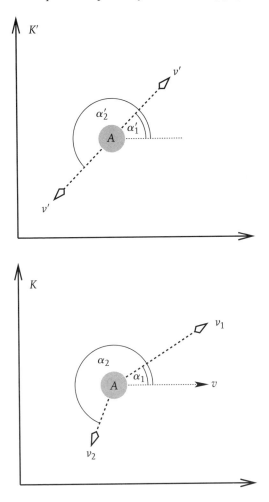

Fig. 3.14 Emission of two photons by a body *A*: in the upper picture as reckoned in the rest system K' of *A*; in the lower picture as reckoned from system *K*, relative to which *A* moves with velocity v in the *x* direction. The lengths of the arrows are proportional to the frequencies. Frequencies and angles change according to Doppler shift and aberration.

measured in K'. This leads to

$$\nu = \nu'\gamma(1 - \beta \cos\alpha'), \tag{3.19}$$

which allows us to determine the frequencies ν_1 and ν_2 measured in K as a function of the angles α'_1 and α'_2 measured in K'. The point being, that the latter two just differ by 180°, so that $\cos\alpha'_1 = -\cos\alpha'_2$. The total energy $\Delta E = h\nu_1 + h\nu_2$ of the photons in K now equals

$$\Delta E = \gamma \Delta E'. \tag{3.20}$$

To be sure, we require that this process obeys the law of energy conservation as reckoned from both systems K and K'. Let E_i and E_f denote the total (kinetic plus inner) energy of A in K before (i) and after (f) the emission process. Let E'_i and E'_f be the corresponding quantities in K'. Then we must have

$$\begin{aligned} E_i &= E_f + \Delta E, \\ E'_i &= E'_f + \Delta E'. \end{aligned} \tag{3.21}$$

By its very definition, the kinetic energy of A in K before emission must equal the difference $E_i - E'_i$. The corresponding equation holds after emission. Subtracting the second from the first equation in (3.21) and using (3.20) yields the following expression for the difference of the kinetic energies before and after emission:

$$\Delta E_{\text{kin}} = \Delta E'(\gamma - 1). \tag{3.22}$$

We now compare this with the general expression (3.17) for kinetic energy. Recall that A's velocity does not change during the emission process. Hence a change of its kinetic energy can only be brought about by a change Δm_0 of its rest mass. According to (3.16), the latter will then cause a change in the kinetic energy of the form $\Delta m_0 c^2(\gamma - 1)$. This is just of the same form as the right-hand side of (3.22). Writing now ΔE_0 instead of $\Delta E'$ for the variation of A's energy in its rest system, we obtain by comparison

$$\Delta E_0 = \Delta m_0 c^2. \tag{3.23}$$

For completeness we remark that instead of energy conservation we could have used momentum conservation to obtain the same result. One may show that A suffers a recoil in K corresponding to a momentum transfer in the x direction of $\Delta p_x = -\gamma(\Delta E'/c^2)v$. Since v stays constant, the rest mass must have diminished by $\Delta m_0 = \Delta E'/c^2$.

This answers the question that Einstein posed in the title of his paper in the affirmative. The rest mass, and hence the inertial mass, depends on the internal energy of the body. An increase of internal energy by an amount ΔE also increases the rest mass by $\Delta E/c^2$. Note that the argument is totally independent of the form in which this energy is stored within the body. Taken together with the already established connection (3.18) between inertial mass and kinetic energy, this leads to the statement that *any* change in energy, whether it be internal or external (kinetic), corresponds to a change in the inertial mass, and vice versa. Note that the rest mass of an extended piece of matter is defined as its mass, as measured in the rest system of its centre of mass. The rest mass receives contributions from all constituents of the body, including their kinetic, potential, chemical, and other energies. For example, the rest mass of a piece of metal will depend on its temperature, since temperature stands for an average of the kinetic energies of the atoms and molecules.

Eventually this leads to an identification of the notions of 'inertial mass' and 'energy', whose physical definitions did not at first depend on each other. Hence, after fixing the otherwise undetermined additive constant for energy such that $E = 0$ for $m = 0$, (3.18) and (3.23) may be summed up in a formula for the total (internal plus kinetic) energy, known to almost everybody:

$$E = mc^2. \tag{3.24}$$

This equation is totally general and applies to all physical systems. Knowing the total energy we can calculate the total inertial mass and vice versa. Note that a separation of mass into rest mass and kinetic energy, like in (3.18), only makes sense for systems which can be assigned a collective velocity, like for movable bodies. Then we may write $m = m_0\gamma$, as in (3.15). Alternatively we may express E in terms of the momentum (3.16) rather than the velocity. This leads to the following relation

$$E^2 = p^2c^2 + m_0^2 c^4. \tag{3.25}$$

This formula is of central importance in Relativistic Quantum Mechanics and Quantum Field Theory and much used in atomic and elementary particle physics. We will come back to this in Sects. 4.1 and 4.3. The relation (3.25) between energy and momentum is the same in all inertial systems. This is a consequence of the fact that under Lorentz transformations both quantities get mixed in a way that is identical to the transformation rules for time and space coordinates. This will be discussed in Sect. 5.9.

On a quantitative level, the most remarkable property of (3.24) is the magnitude of the factor c^2 that converts masses into energies. Expressed in units of square metres per square seconds, it is roughly given by 10^{17}! This is the number you have to use in order to convert a mass in units of kilograms into an energy in units of joules. For example, the kinetic energy of a luxury limousine of a mass of two tonnes at a speed of 200 kilometres per hour corresponds to a tiny mass of only $3.4 \cdot 10^{-8}$ grams. This enormous magnitude of c^2 becomes important in nuclear physics, where it explains the high energy yield of nuclear fission and nuclear fusion. We will come back to this in Sect. 4.2.

3.10 Electrodynamics: Invariance of Maxwell's equations

We have seen how SR implies certain changes to be made in the formalism of mechanics. These changes can be understood as consequences of the requirement that the equations of motions be invariant under Lorentz transformations (instead of Galilei transformations). This, in fact, would have been the shortest, though rather formal, route to special-relativistic mechanics. This axiomatic approach is the most commonly adopted one in modern

Invariance of Maxwell's equations

textbooks. Now, in contrast to Newtonian mechanics, Maxwell's equations are not invariant under Galilei transformations. This has for a long time been the mathematical origin of the belief that electrodynamics would be incompatible with the principle of relativity. But we now understand that this is based on the hidden assumption that the change of inertial systems, which are physically defined, is mathematically implemented by Galilei transformations. For a long time this was thought to be self-evident. But, as Einstein showed, this is based on a prejudice concerning the physical meaning of space-time measurements. The physically correct implementation of the principle of relativity must be via Lorentz transformations.

Now, the important point is that Maxwell's equations are already Lorentz invariant. Hence, according to Einstein, they already do satisfy the principle of relativity. It is interesting to note that this mathematical result per se was established before Einstein by Lorentz (1904) and Poincaré (1905). This, by the way, is the reason the Lorentz transformations carry their name. In fact, similar wave-propagation equations to those that appear in Maxwell's theory where already shown to be Lorentz invariant by the mathematician Woldemar Voigt (1850–1919) in 1887. But nobody before Einstein connected these results to the principle of relativity. In particular nobody took the transformation of the time coordinate as anything else but a formal manipulation, disconnected from any physical notion of time.

That Maxwell's equations are Lorentz invariant means the following. Let K and K' be two inertial systems whose coordinates are (x, y, z, t) and (x', y', z', t'), respectively. Then there is a unique Lorentz transformation (including rotations and translations in the general case), L, that transforms the first into the second set of coordinates. Let $\vec{E}$ and $\vec{B}$ be electric and magnetic fields, measured with yardsticks and clocks in K and expressed as functions of the coordinates of K. We assume these fields to satisfy Maxwell's equations in K. Then there exists a unique transformation, $(\vec{E}, \vec{B}) \rightarrow (\vec{E}', \vec{B}')$, such that the new fields satisfy Maxwell's equations in K'. Here $\vec{E}'$ and $\vec{B}'$ are the electric and magnetic fields, measured with yardsticks and clocks in K' and expressed as functions of the coordinates in K'.

As an example we wish to write down the transformation formulae for the electromagnetic fields in case K' moves relative to K at speed v in the x direction, where, as usual, both systems coincide at $t = t' = 0$. Then the space-time coordinates transform as in (3.3) and the fields as follows (we express the fields in K as functions of the fields in K'):

$$E_x = E'_x, \ E_y = \gamma(E'_y + vB'_z), \quad E_z = \gamma(E'_z - vB'_y),$$
$$B_x = B'_x, \ B_y = \gamma(B'_y - \frac{v}{c^2}E'_z), \ B_z = \gamma(B'_z + \frac{v}{c^2}E'_y). \quad (3.26)$$

A remarkable property of these equations is that electric and magnetic fields mutually transform into each other. Hence the split between electric and magnetic components of the electromagnetic field depends on the observer's state of motion. No absolute distinction between 'electric' and 'magnetic' exist anymore. Let, for example, the field in K' be purely electric ($\vec{B}' = \vec{0}$) and pointing in the z' direction with constant strength E'_z. An observer at rest in K will then not only measure an electric field in the z direction of enhanced strength $\gamma E'_z$, but perpendicular to it also a constant magnetic field in the y direction of strength $B_y = -\gamma v E'_z / c^2$.

This is precisely the effect which lifts the apparent dichotomy in the explanation of electric induction, which Einstein emphasized so much right at the beginning of his original paper on SR. Once Maxwell's equations are recognized to be applicable in all inertial systems, we can now use them either in the rest system of the magnet or the rest system of the moving conductor (wire). The absolute split between electric and magnetic, which existed as long as one believed the ether system to be the only one where Maxwell's equations apply, is now gone! Let us again take a look at **Fig. 3.1**. Let the y direction point upward within the plane of the paper and the z direction toward the reader, perpendicular to the plane of the paper. The rest system K' of the magnet then contains

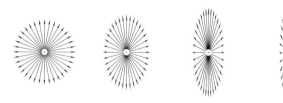

Fig. 3.15 Coulomb field of a point charge in various states of motion. From left to right: at rest and with velocities 0.5, 0.7, and 0.8 in units of c. Depicted are the vectors of the electric field, evaluted on the surface of the small sphere of constant radius (measured in the laboratory system K) about the central charge.

a constant magnetic field in the z' direction of strength B'_z in the region marked by the $\otimes$ symbols. Now we switch to the rest system K of the conductor. According to (3.26), it contains an electric field in the y direction of strength $E_y = \gamma v B'_z$. This is precisely the field that the law of induction would predict by solving Maxwell's equations in K. But Lorentz invariance of Maxwell's equations spares us the trouble of doing this calculation. We can predict the result straightaway by doing the much easier job of solving Maxwell's equations in K' and *then* transform the result to K following the rules (3.26). This shows that Lorentz invariance can also be put to great practical use.

Finally we consider the electric field of a (positive) point charge. Let K' be its rest system with the charge at the origin. In K' the electric field is given by attaching to each point of space a radial outward pointing vector, whose length is inversely proportional to the square of the distance to the charge. Depicting a few of these vectors at constant angular separation and fixed radial distance leads to the first picture in **Fig. 3.15**. We are interested in the analogue of this picture in K, relative to which the charge moves with velocity v in the x direction. Transforming the first picture according to (3.26) leads to an unchanged horizontal component of the electric field, whereas the transverse component gets multiplied

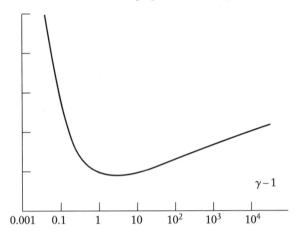

Fig. 3.16 Ionization power of a charged particle as a function of its kinetic energy.

(enhanced) by the factor γ. The additional magnetic field does not interest us now. But this is not yet the sought-for analogue because the footpoints of these vectors now no longer lie on a sphere of radius r. Length contraction squashes the original sphere horizontally by a factor of $1/\gamma$, so that the field vectors we have obtained now are those evaluated on that ellipsoid. Correcting for this, i.e. evaluating the field truly on a sphere in K, we get an additional weakening of the horizontal components by γ^{-2} due to the electric field's $1/r^2$ fall-off. Hence, in total, we get a field whose vertical components are enhanced by a factor γ and whose horizontal components are weakened by a factor of γ^{-2}. This we depict in **Fig. 3.15** for the velocities $v = 0.5c$, $v = 0.7c$, and $v = 0.8c$. The enhancement of the vertical component has direct experimental consequences. It is responsible for the growing of the ionization power of very fast moving electric particles in media. Without the relativistic effect one would expect a decay with the particle's velocity, due to the shorter time it now spends in the vicinity of the matter atom. The qualitative behaviour is plotted in **Fig. 3.16**. On the horizontal axis we plotted $\gamma - 1$, i.e. the kinetic energy in units of the rest energy m_0c^2, and on the vertical axis the energy transfer per unit path length, which is a direct measure for the ionization power. There is an obvious turnaround at about $\gamma = 3$, which corresponds to a speed of about 95% of the speed of light.

·4·
Further consequences and applications of Special Relativity

4.1 Atomic physics

Atomic physics is essentially based on Quantum Mechanics. Quantum Mechanics makes essential use of the variables that describe positions and momenta (rather than velocities). The starting point for the determination of atomic energy spectra is the quantum mechanical transcription of the energy conservation equation, which in the 'non relativistic' situation is the famous 'Schrödinger equation'. Schrödinger's equation is based on the Newtonian relation between energy and momentum of a particle of mass m_0, given by $E = p^2/2m_0$. In special relativistic mechanics, this relation is replaced by (3.25), which approximates the former for momenta that are small compared to mc, but also implies significant deviations from it for larger momenta. In Quantum Mechanics these deviations give rise to 'relativistic corrections' to the 'non-relativistic' (i.e. based on the Schrödinger equation) energy levels. A second 'relativistic' correction stems from the transformation rule (3.26), which predicts a magnetic field in the rest system of the electron that moves in the electric field of the atomic nucleus. This magnetic field interacts with the magnetic moment of the electron, thereby producing extra energy contributions.

The simplest and lightest atom is that of hydrogen. It merely consists of a single positively charged proton (the nucleus) and a single electron, of equal and opposite charge, orbiting the nucleus. Here the electron reaches velocities up to 0.7% of the velocity of light. Relativistic corrections, which are of the order of v^2/c^2, are therefore expected on the scale of $5 \cdot 10^{-5}$. Well-developed techniques in quantum mechanical perturbation theory allow us to calculate these corrections. In leading order the sum of both effects mentioned above gives the following 'relativistic' corrections to the energy levels of hydrogen, labelled by the principal quantum number n and total (orbital plus spin) angular momentum j:

$$\Delta E_{nj} = E_1 \frac{\alpha^2}{n^3} \left(\frac{3}{4n} - \frac{1}{j + \frac{1}{2}} \right). \qquad (4.1)$$

Here $E_1 = 13.61$ eV is the unperturbed energy of the ground state (the tightest bound state) and $\alpha \approx 1/137$ is the so-called 'fine-structure constant'. (eV denote the energy unit 'electron volts'.)

These corrections are called the 'fine structure' of the energy levels of hydrogen. Next to the principal quantum number n they also explicitly depend on the quantum number j for angular momentum, in contrast to the unperturbed levels which merely depend on n. This means that different states of the atom, which according to Schrödinger's equation have the same energy (so-called energetically 'degenerate' states), now turn out to be, in fact, energetically different. This is sketched in **Fig. 4.1** for the three lowest lying states, $n = 1, 2, 3$, of hydrogen. To the left of the vertical line we have put the uncorrected levels, as predicted by Schrödinger's equation. To the right we see the corrected levels. Note that one level on the left may consist of several degenerate atomic states. Hence one level may, after correction, split into several levels, depending on its value for total angular momentum j. For technical reasons of presentation the distances between the levels are not drawn to scale. In particular, we strongly contracted the vertical distances between levels of different n so that the corrections appear strongly exaggerated. It is apparent that all levels shown are lowered and degeneracies between levels of equal n but different j are lifted. The numbers written just below the lowered levels denote $(\Delta E_{nj}/E_1) \cdot 10^6$, that is, the energy shift in units of a millionth of the uncorrected ground-state energy.

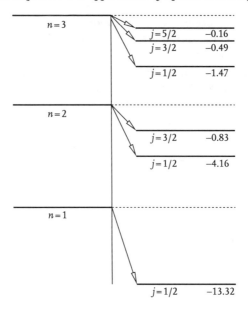

Fig. 4.1 Lowest energy levels of the hydrogen atom.

Finally we note for completeness that there are further, though still smaller, corrections to the energy levels of atoms. These, on one hand, stem from effects of Quantum Electrodynamics (Lamb shift), and on the other hand, from taking into account the interaction of the nucleus with the magnetic field produced by the moving electron. The latter gives rise to the so-called 'hyperfine structure'.

4.2 Nuclear physics

Atomic nuclei are made of electrically positively charged protons and electrically neutral neutrons. They are almost of the same mass—in fact, the neutron is heavier than the proton by one part in a million. Together they are called nucleons. The number A of protons plus neutrons is the 'mass number', whereas the number of protons alone, Z, is called the 'atomic number'. It is the latter number that determines the position of the corresponding element in the periodic table. The nucleus itself is named after its corresponding chemical element. Nuclei with equal Z but different A, i.e. the same number of protons but different number of neutrons, are called 'isotopes'. To precisely characterize a particular isotope of an element E one attaches the mass and atomic numbers in the form ^A_ZE. For example, standard helium, consisting of two nucleons of each sort, is denoted by $^4_2$He.

A configuration of protons and neutrons could not stay together if there were no force acting against the electrostatic repulsion of the like charged protons. This force is called the strong or nuclear force. That there exists a stable bound state means that the energy of the compound system is less than the energy of its parts in isolation. The difference is called the binding energy. Atomic physicists know binding energies per electron up to some two-digit electron volts (compare the 13.6 eV for the ground state of hydrogen, mentioned above). Binding energies in nuclear physics, however, lie on the order of a few MeV (MeV, a million electron volts) per nucleon, so roughly a hundred-thousand times larger.

According to the universal relation $E = mc^2$ this binding energy, too, corresponds to a mass. This implies that the bound configuration has a mass that is less than the sum of the masses of its constituents. This is generally known as the 'mass defect' of bound systems. Due to the large conversion factor, c^2, mass defects are often too small to be of any significance. This is different for nuclear binding energies, which are sufficiently strong to cause measurable mass defects of nuclei. Here one has to recall that masses of nuclei can, in fact, be determined to better than one part in a million, whereas typical mass defects are of the order of a few per cent of a nucleon mass. For example, the helium nucleus, also known as the α-particle, which is composed of two protons and two neutrons, has a mass that is less than the sum of two proton and two neutron masses by three per cent of a nucleon mass. This corresponds to a total binding energy of almost 30 MeV, that is, 7 MeV per nucleon.

The binding energy per nucleon grows with the mass number of the nucleus and reaches a maximum of almost 8.8 MeV for the

90   Consequences and applications of Special Relativity

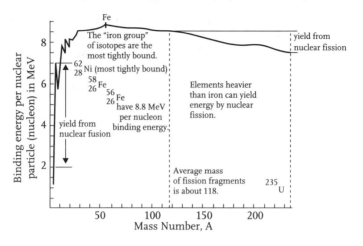

Fig. 4.2 Binding energy per nucleon as a function of the mass number.

most tightly bound nuclei. Ordered according to increasing binding energy, the three most tightly bound ones are given by the following iron and nickel isotopes: $^{56}_{26}$Fe, $^{58}_{26}$Fe, and $^{62}_{28}$Ni. For still higher mass numbers the binding energies per nucleon fall off gently. This is plotted in **Fig. 4.2**. Roughly speaking this means that up to mass number $A \approx 60$ energy is gained by nuclear fusion, that is composition of lighter nuclei into heavier ones. On the other hand, beyond $A \approx 60$, energy is gained by fission, that is decomposition of heavy nuclei into lighter ones. This, too, is depicted in **Fig. 4.2**. Stars gain their radiation energy from nuclear fusion. Our Sun, for example, radiates approximately $4 \cdot 10^{26}$ joules per second. This corresponds to an equivalent of 4.4 million tonnes which the Sun loses in mass every second. That mass is taken from the nuclear binding energy and radiated away through fairly complicated processes, in the course of which four hydrogen nuclei $^1_1$H (protons) eventually combine into one helium nucleus $^4_2$He, leaving two positrons (the antiparticle of the electron) and two neutrinos. Each of these processes sets free the energy equivalent of approximately the mass defect of $^4_2$He, that is 30 MeV. Hence one needs a rate of 10^{38} such processes per second to fuel the current radiation power of the Sun. This costs the Sun $4 \cdot 10^{38}$ hydrogen nuclei per second. Given that roughly 70% of the Sun's mass is provided by hydrogen and that the Sun's total mass is $2 \cdot 10^{30}$ kg, we can estimate the number of hydrogen nuclei in the Sun to be of the order of 10^{57}. If all of that could be burned into helium, the Sun would be able to keep up its present radiation power for at most 70 billion ($7 \cdot 10^{10}$) years. A more detailed analysis shows that the future life of our Sun will be shorter than this by slightly more than a factor of ten. This is partially due to the fact that the Sun will not completely burn its hydrogen fuel, and partly due to the fact that the burning rate will increase toward the end. In fact, the Sun is expected to shine for another 6 billion years from now. Likewise it also follows that the hydrogen to helium fusion process is sufficiently effective to have let the Sun shine at its current radiation power in the last 4–5 billions years without burning out. How this might be possible was considered a big mystery up to and far into the 20th century (see, e.g. the lucid account in [19]).

Note that it is not SR that physically explains why the binding energies of nuclei are as big as they are. Rather this is done by the theory of the strong interaction—Quantum Chromodynamics. But it was, and still is, the special relativistic mass defect that gave the first and effective means to determine the binding energy without knowing anything about such a theory.

Another early application of $E = mc^2$ was made after nuclear fission was found in 1938. The natural question then was how much energy would be released in the fission of a single uranium nucleus. It was realized that the mass of the uranium nucleus was bigger than the sum of the masses of the fission products, barium and krypton. Already in 1939 this was correctly interpreted as an (inverse) mass defect. Einstein's formula then immediately led to an energy release per uranium nucleus of roughly 200 MeV. This made it all too clear what an enormous energy reservoir uranium was. The issue of August 15th 1939 of a big Berlin newspaper contained an article by a young German physicist, Siegfried Flügge, who illustrated this magnitude in energy gain in the following manner: if all the uranium contained in a cubic metre of uranium oxide (U_3O_8) underwent fission, it would supply sufficient energy to lift up one cubic kilometre of water to an altitude of 27 kilometres above ground. More drastically: it would be sufficient to 'hurl' the Berlin Wannsee (a fairly big lake near Berlin) into the stratosphere!

4.3 Elementary particle physics

The by far largest and also most important realm of applications for SR is undoubtedly high-energy elementary particle physics. Its methods and concepts would be unthinkable without SR. Here the consequences of SR reach from simple corrections concerning the dynamics of fast moving particles to profound revisions concerning our very notion of 'matter'. Let us start with two examples of the first kind.

- The relativistic relation (3.16) between momentum and velocity leads to an increase of the centrifugal force with orbital velocity that is stronger by a factor of γ than that derived from Newtonian mechanics. This needs to be taken into account in the design of particle accelerators. As an example consider a particle of rest mass m_0 and electric charge e that is injected at velocity v perpendicular to the field lines of a constant magnetic field. The special relativistic equations of motion predict that the particle will move on a circular orbit of radius

$$R = \gamma \frac{m_0 v}{eB}. \qquad (4.2)$$

In contrast, the Newtonian equations of motion predict the smaller radius that one obtains from (4.2) by dropping the factor γ. If one were to design the radius of curvature of the accelerator's vacuum tubes according to the Newtonian prediction, the particle beam would hopelessly 'understeer' and run straightaway into the tube's outer boundary. As an

Elementary particle physics 93

extreme case we mention the Tevatron ring at Fermilab, in which protons reach energies of 1 TeV $= 10^{12}$ eV. The proton's rest energy, $m_0 c^2$, is just below 1 GeV $= 10^9$ eV, so that γ-factors of a thousand are reached!

- Time dilation and length contraction are everyday occurrences in particle physics. A classic example is given by muons (or μ-mesons). These are particles of the same electric charge as electrons or positrons but 207 times heavier, corresponding to a rest energy of 106 MeV. However, muons are highly unstable particles: they rapidly decay with a mean lifetime τ_μ of only 2.2 microseconds (10^{-6} s) into an electron or positron, a μ-neutrino, and an e-antineutrino. Muons are, for example, produced when highly energetic cosmic rays collide with atoms in the upper layers of the Earth's atmosphere, typically at 15–20 kilometres altitude. But even if the muon travelled with the speed of light, it should not get much further than $c \cdot \tau_\mu = 660$ metres before decay. Certainly it should not be able to reach the surface of the earth. But here experimentalists still detect a muon flux of about one muon per minute per square centimetre, which corresponds to a significant fraction of all muons. How can this be? The explanation is twofold, depending on whether one takes the point of view of an observer on Earth or in the rest system of the muon. Relative to the observer on Earth the muon-clock is slow by a factor of $1/\gamma$. But only according to the latter clock does the muon decay with mean lifetime $\tau_\mu = 2.2 \cdot 10^{-6}$. According to the clock on Earth, the mean lifetime appears stretched by a factor γ. So, according to SR, the observer on Earth reckons the average atmospheric penetration depth of muons to be $\gamma c \tau_\mu = \gamma \cdot 660$ m. Typical energies for the muons are of the order of 20 GeV, which makes a γ-factor of almost 200. This fully explains the muon detection rate on the Earth. Alternatively, the same situation may be described in the rest system of the muon. Now the mean lifetime is not dilated. Instead the muon's distance to the surface of the Earth suffers a length contraction by a factor of

$1/\gamma$. The result concerning the fraction of muons that reaches the Earth is the same. Fully analogous considerations hold for lifetimes and ranges of particles in accelerators, where, as already mentioned, γ-factors of about 1000 are reached.

We now wish to touch upon a very fundamental point, where the impact of SR drastically changes our very concept of 'matter'. The naive interpretation of the term 'elementary particle' is that of an everlasting piece of matter—an object—that cannot be decomposed by physical means and that obeys simple laws of motion. The assumption of the existence of such elementary objects is usually connected with the hope to be able—eventually—to reduce the complex phenomena in nature to simple laws that govern the motions of such objects and the ways in which they may combine into larger ones. Such a reductionist programme is attractive for its logical clarity and the potentially high explanatory power. Subscribing to it means to set out and (1) identify these objects, and (2) find their laws of motion and how they interact with each other.

Special Relativity irreversibly puts an end to this programme, at least if understood in the strict and somewhat naive way just outlined. This is because SR denies the existence of everlasting stable objects. The equivalence of mass and energy, expressed in $E = mc^2$, allows various transformations between particles of different kinds. It even allows particles to be created out of purely unstructured energy, like the kinetic energy of an already existing particle, or some radiation energy. This is almost as if they appeared out of nothing, as long as the energy balance and some other existing conservation laws, which we did not mention so far, permit the deal.

At this point equation (3.25) plays a crucial rôle. Since it is *quadratic* in E (unlike its Newtonian counterpart) it allows *two* solutions for the energy, given the particle's momentum and mass. The negative-energy solution corresponds to the antiparticle for the particle that is described by the positive-energy solution. The existence of antiparticles can be seen as direct consequence of combining SR with quantum physics. The first experimental evidence in this direction came with the detection of the positron in 1931, after it had been predicted by Paul Dirac (1902–1984) in 1928 on the basis of his Lorentz-invariant generalization of Schrödinger's equation. (Actually Dirac first thought that his 'other solution' should be identified with the already known proton. But being 1836 times heavier than the electron, that was incompatible with his equation, which demands particle and antiparticle to be of equal mass.)

To be sure, such mutations between particles of all sorts cannot be just arbitrary. There are a number of conservation laws that all dynamical processes involved must respect. Next to energy these concern momentum, electric charge, and also some other charges as well. But there is still a great variety of 'channels' along which mutation processes can occur. Hence, in principle, the concept of everlasting elementary objects, first conceived by the ancient atomists in the 5th century BC, seems irreversibly gone.

As an example of such a mutation between different forms of matter we take a look at **Fig. 4.3**. It shows the twofold transformation of a highly energetic photon ('γ-ray') into an electron–positron pair in a bubble chamber. Since bubble chambers only show the

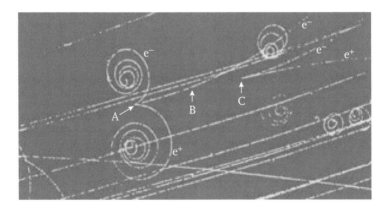

Fig. 4.3 Bubble chamber traces of a twofold production of electron–positron pairs by a γ-ray.

traces of electrically charged particles, the photon cannot be seen. It enters the chamber from the lower left direction and remains invisible until it produces a pair consisting of one electron (e^-) and one positron (e^+) at the position marked A. Perpendicular to the plane of the picture and pointing towards the reader is a magnetic field that forces negatively charged particles onto left-, and positively charged particles onto right-hand bends. The large curvatures (small radii) of the upper and lower particle traces emerging from A indicate small velocities, according to (4.2). The third trace that emerges from A and runs to B on a light left-hand bend corresponds to a much faster electron that already existed at A and was hit very hard by the incoming photon. This now highly energetic electron in turn emits a photon at B, which, invisibly, runs to C where it produces yet another electron–positron pair. This latter e^-e^+ pair is more energetic than the first one, as one easily infers from the smaller curvatures of their traces.

Today's concept of 'matter' differs quite drastically from the more naive ones that prevailed upto and into the 20th century. It is based on more abstract concepts of relativistic Quantum Field Theory. In particular, the notion of a 'particle' gets absorbed into the notion of a 'quantum field', which is a structure that extends throughout space-time. The fields are the fundamental entities of the theory, whereas particles correspond to some (quantized) excitations of these fields, which do have certain spatially localizable properties. A separation of all particles into those which are 'elementary' and those which are not is neither necessary nor does it seem to be natural anymore. There is also nothing that would correspond to the classical idea of the vacuum, i.e. a space devoid of any matter. Quantum fields are always there and cannot just be set to 'zero'. They make themselves felt anywhere at any time through typical quantum fluctuations of physical quantities, like energy. Although their contribution in absolute value to the total energy in a fixed volume cannot be calculated in a meaningful way (the mathematical expressions for these fluctuation contributions formally diverge, even if the considered volume is finite), they do give rise to physically measurable effects concerning the value of energy *differences*, which can be calculated. Recently this 'vacuum fluctuation energy' became a fundamental issue in connection with the measured accelerated expansion of the Universe. This is because such an accelerated expansion can only be driven by very special kinds of gravitating matter, whose internal pressure needs to be exceptionally high and negative, as compared to its energy density. No known matter on Earth shows such a strange behaviour. But the energy–pressure relation that the theory predicts for the vacuum fluctuations of quantum fields does. The unfortunate fact is that even though the energy density, ρ, and the pressure, p, share the right relation, namely $p = -\rho$, both quantities cannot be predicted in absolute value, as already mentioned. This is because, as it stands, the theory predicts quantum fluctuations to exist at arbitrarily high frequency scales in any volume, however small, which sum to an infinite energy contribution. In other words, the value of the energy density due to vacuum fluctuations comes out to be infinite, which, physically speaking, is sheer nonsense. It is generally expected that the theory ceases to be correct at the smallest scales, certainly beyond the Planck scales (2.6), (2.7), where quantum gravity effects are expected to provide a dynamical regulating mechanism that damps out the highest frequency contributions (higher than the 'Planck frequency' $\nu_P = 1/t_P$). But even if in the above calculation one cuts off (by hand) all fluctuations corresponding to energies (= frequency $\cdot h$) higher than, say, 100 GeV (Fermi scale), which is well within the reach of modern accelerators and at which modern theories are also well tested, one still obtains a quantum-fluctuation energy density that is above the cosmologically measured one by 52 orders of magnitude! So presently it seems hopeless to 'explain' the Universe's accelerated expansion as being driven by vacuum fluctuations.

Further fundamental consequences of SR in Quantum Field Theory are the so-called 'Spin-Statistics Theorem' and the 'PCT Theorem'. The first entails a strict correlation between the spin (intrinsic angular momentum in units of $\hbar$) of the particle and the type of statistics it obeys. The second requires the combination of certain operations to be a fundamental symmetry, even though the individual operations are not. The operations are: space reflection (P), charge conjugation (C), and time inversion (T).

4.4 Daily physics: navigational systems

Navigational systems serve to determine one's position on Earth. This is done through the determination of the distances to several known reference points. For example, when at sea, it is usually sufficient to know the distances to two coastal cities. These determine two circles, one around each city, with generally two intersection points, one at sea, the other on land. If both intersection points lie at sea and one has no other hint as to which might be the right one, a further distance to a third reference point is needed. In three-dimensional space two distances determine two spheres which generically intersect in a circle. A third distance then generically selects two points, which is often sufficient; for example, if one of the points is not on the surface of the Earth.

In modern navigational systems, like the the GPS (global positioning system) of the US Department of Defense, or its Russian counterpart GLONASS (global navigation satellite system), the 'reference points' are satellites that orbit the Earth on accurately known trajectories. The GPS space segment consists of at least 24 satellites in six almost circular orbits whose radii are all close to 20 thousand kilometres. The orbits have a relative inclination of 56 degrees, so that at least four satellites can be 'seen' from any (obstruction free) location on Earth at any time. The distances to the satellites are determined via travel times of electromagnetic signals, sent out by the satellites and collected by receivers carried by the users. These times are converted into distances by multiplying them by the speed of light. At this point it is absolutely crucial that the speed of light does not depend on the emitter's, i.e. the satellite's, state of motion.

In one nanosecond (10^{-9}s) light travels a distance of 30 centimetres. Hence the travel times must be determined to an accuracy

Daily physics: navigational systems

of 100 nanoseconds in order to achieve a localization precision of 30 metres. Suppose the clocks in the satellites can be adjusted at most every 24 hours. Then its daily (= 86 400 s) deviation must not exceed 100 nanoseconds. In other words, its relative error tolerance must be better than or equal to $100 \cdot 10^{-9}/86\,400 \approx 10^{-12}$. In fact, the atomic clocks mounted on the satellites, and certainly those used on Earth, beat this limit by one or more orders of magnitude.

In order to measure one-way travel times, it is necessary to relate all clock readings to a globally defined 'time', which comprises the satellite clocks as well as all the GPS clocks on Earth. This global time corrects for the systematic differences in the clock rates. There are many effects which contribute to such relative changes of clock rates. Clearly, the ones of interest to us here are the special relativistic effects, though they are not the dominant ones. They are also intimately related, and of comparable order, to other effects caused by the Earth's gravitational field. Even though, in principle, the latter need to be described in the context of General Relativity, which is outside the scope of this book, we can give a somewhat simplified description.

Relativistic effects dominantly concern a systematic deviation of clock rates between the group of clocks stationed on the surface of the Earth on one hand, and the group of clocks mounted on satellites on the other. The latter suffer a time dilation according to SR. But due to the Earth's gravitational field there is another effect that works in the opposite direction. This is because the two groups of clocks are located in regions of different gravitational potential. Let ϕ_E and ϕ_S denote the gravitational potentials on the surface of the Earth (sea level, say) and at the altitude of the satellite orbits, respectively. Let Δt_E and Δt_S be the intervals by which the individual readings of the clocks on Earth and on board the satellites, respectively, proceed if the global time advances by one unit. Then General Relativity predicts that the clock at the lower gravitational potential advances less than the clock at higher potentials. Their relative deviation according to this effect is denoted by Σ_{Grav}. More precisely, in leading order we have

$$\sum_{\text{Grav}} = \frac{\Delta t_S - \Delta t_E}{\Delta t_E} = \frac{\phi_S - \phi_E}{c^2} = \frac{R_G}{R_E} - \frac{R_G}{R_S}. \quad (4.3)$$

Here R_E is the Earth's radius and R_S the radius of the circular satellite orbit. In the last step we also used the fact that the gravitational potential above the Earth's surface (i.e. for $r > R_E$) is of the form $\phi(r) = -GM_E/r$, where G is Newton's constant and M_E the mass of the Earth. As an abbreviation we also introduced the quantity $R_G = GM_E/c^2$. It re-expresses the mass M_E in terms of a length, the so-called 'gravitational radius' of M_E, just by multiplying it with the constant G/c^2, the physical dimension of which is length-over-mass. For the Earth this gravitational radius corresponds to 4.4 millimetres. (Its meaning in General Relativity is the following: if you want to turn the Earth, or any other body, into a black hole, you have to compress it into a volume whose diameter is of the order of its gravitational radius; cf. Sect. 4·6.) Since $R_S > R_E$, the right-hand side of (4.3) is positive. This means that the gravitational effect wants the satellite clocks to run ahead of the clocks stationed on Earth.

The relative deviation in clock readings according to time dilation is denoted by Σ_{SR}. According to (3.4) it is given (in leading order of v^2/c^2) by

$$\sum_{\text{SRT}} = \frac{\Delta t_S - \Delta t_E}{\Delta t_E} = -\frac{v^2}{2c^2} = -\frac{R_G}{2R_S}, \quad (4.4)$$

where v is the velocity of the satellites relative to the surface of the Earth. In the last step we used the law of energy conservation for the satellite to eliminate v in favour of the gravitational potential. This makes (4.4) formally similar to (4.3). The total effect is given by the sum of these two effects:

$$\sum = \sum_{\text{Grav}} + \sum_{\text{SRT}} = \frac{\Delta t_S - \Delta t_E}{\Delta t_E} = \frac{R_G}{R_E} - \frac{3}{2}\frac{R_G}{R_S}. \quad (4.5)$$

Daily physics: navigational systems 101

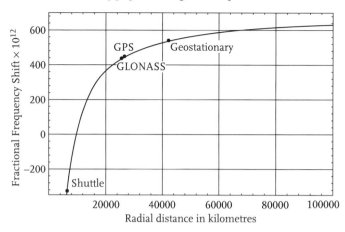

Fig. 4.4 Total relative deviation between satellite clocks and clocks on the surface of the Earth, due to relativistic effects, as a function of the satellite's orbital radius.

We see that the special relativistic time-dilation effect dominates for satellite orbits below 1.5 times the radius of the Earth; that is, 9500 kilometres of radius or 3180 kilometres of altitude above the Earth's surface. Hence, for low-lying orbits, the orbiting clocks lag behind clocks on Earth. This is, e.g., the case for the Space Shuttle.

In contrast, for orbits above 1.5 R_E, the gravitational effect dominates and clocks on Earth lag behind satellite clocks. This is the case for the GPS and GLONASS systems. **Figure 4.4** depicts the total effect, Σ, as a function of the circular orbit's radius R_S. For the GPS we have $R_S = 4.2\, R_E = 2.66 \cdot 10^7$ m, so that $\Sigma_{SRT} = -0.83 \cdot 10^{-10}$ and $\Sigma_{Grav} = 5.25 \cdot 10^{-10}$. These deviations lie above the sensitivities of modern atomic clocks by about five orders of magnitude.

Failing to correct a relative deviation between the satellite clocks on one hand, and the clocks on Earth on the other, would within six hours accumulate to deviations of five millionths of a second. In turn, this would result in positioning errors of about a kilometre. This would render the navigational system totally useless for, e.g. daily traffic navigation. This example clearly shows that relativistic effects are already part of our daily life.

4.5 Science fiction: travel to distant stars?

We have seen above that time dilation allows decaying particles to travel distances much larger than the distance light can reach within the particle's lifetime. This sounds paradoxical at first, as if the particle was claimed to travel faster than light. But this is clearly not the case. There is no contradiction, because the lifetime refers to the time measured in the rest system, K', of the particle, whereas the time that one multiplies the velocity with in order to calculate the distance is measured in the laboratory system, K. What is true for particles and their lifetimes also applies to living organisms, humans in particular. Sufficiently fast spacecraft should therefore enable the crew to explore regions in the Universe more remote than, say, 100 light years. Can this really be true?

Let us consider a concrete example. Suppose we built a spaceship whose engines can provide sufficient thrust to uphold a constant acceleration of $a' = 10 \text{ m/s}^2$. This acceleration is just the same as that due to gravity on the surface of the Earth. The crew should therefore feel perfectly comfortable. More precisely, the quantity a' refers to the acceleration measured in the instantaneous rest system of the spaceship. Now consider a particular moment at which K' is the instantaneous rest system. During the time interval dt' (measured in K') the spaceship acquires a velocity increment (also measured in K') of $dv' = a' \, dt'$. Let K be the rest system of the Earth (which here we may treat as an inertial system). As measured in K it is impossible that the spaceship accelerates at a constant rate, since this would imply that it reaches superluminal speeds within a finite time, an impossibility according to SR. In fact, measured in K, dt' corresponds to the larger time interval $dt = \gamma \, dt'$ and dv' to the smaller velocity increment $dv = \gamma^{-2} \, dv'$. This last equation is obtained by adding to the instantaneous velocity $v(t)$ the increment dv' according to the rule (3.6), keeping only the terms linear in dv'.

Hence the acceleration of the spaceship relative to K is given by $a = dv/dt = \gamma^{-3}(v)\, a'$, which decreases with time (as it must do), since v increases and a' is constant. Using elementary calculus, this equation can now be integrated once to give the velocity v as a function of t, and once more to give the position x also as a function of t. The solution, $v(t)$, can then be used to integrate $dt' = dt/\gamma(v(t))$, which yields t' as a function of t, or vice versa.

We skip the details of the calculation and directly jump to the solution. We are interested in the following analytic expressions: (1) the travelled distance (measured in K) as a function of the time t' that has passed in the spaceship, (2) time t of K as a function of time t', and (3) the γ-factor as a function of x. We let the spaceship start at time $t = 0$ at position $x = 0$ with initial velocity $v = 0$. The solution contains the parameter a' in the combination c/a', which has the physical dimension of time and is approximately given by one year. (The precise value is 0.95 years. In order to make c/a' exactly one year, one chooses $a' = 9.5$ m/s^2.) Hence the analytic expressions assume their simplest form if we agree to understand the times t and t' as being measured in years and the distance x in light years. In this way we can treat these parameters as dimensionless. The solution now reads as follows:

$$x = \cosh(t') - 1, \qquad t = \sinh(t'), \qquad \gamma = x + 1. \qquad (4.6)$$

Here cosh and sinh are the functions called hyperbolic cosine and hyperbolic sine, respectively. They bear the following simple relations to the probably more familiar exponential function:

$$\cosh(t') = \tfrac{1}{2}(e^{t'} + e^{-t'}), \qquad \sinh(t') = \tfrac{1}{2}(e^{t'} - e^{-t'}). \qquad (4.7)$$

For large arguments the hyperbolic cosine and sine essentially grow exponentially. Hence (4.6) shows that after a few years the travelled distance x grows essentially exponentially in time t' that one measures on board the spaceship! In contrast, x as a function of time t measured on Earth grows approximately linearly after a few years, because then the spaceship travels practically at the constant speed of light. Note in particular the strong growth of γ, which is linear

Table 4.1

Destination	Distance [ly]	Duration t' [y]	Duration t [y]	γ
Edge of solar system	5.5 l-hours	13 days	13 days	1.0006
Star Proxima Centauri	4.22	**2.33**	5.12	5.22
Star Vega	26	**3.98**	26.98	27
Our galactic centre	$2.6 \cdot 10^4$	**10.86**	$2.6 \cdot 10^4$	$2.6 \cdot 10^4$
Andromeda galaxy	$2.9 \cdot 10^6$	**15.57**	$2.9 \cdot 10^6$	$2.9 \cdot 10^6$
Virgo cluster	$6 \cdot 10^7$	**18.60**	$6 \cdot 10^7$	$6 \cdot 10^7$
Remotest quasar	$1.3 \cdot 10^{10}$	**23.98**	$1.3 \cdot 10^{10}$	$1.3 \cdot 10^{10}$

in x and hence asymptotically exponential in t'. To illustrate these results **Table 4.1** lists some astronomical destinations, their distances from Earth in light years (ly) ignore the travel times t and t' in years (y), and the γ-factor that is reached at the destination.

The results for t', highlighted in bold face, seem to indicate that in less than 25 years we could quite comfortably cruise right through the entire known Universe. Unfortunately this conclusion does not survive closer scrutiny. This is not to say that there is anything wrong with what we have said so far, but it is definitely not the whole story. For example: consider a journey to our closest neighbouring star, Proxima Centauri. Just before we reach the destination, our space ship moves at $\gamma = 5$ (corresponding to 98% of the speed of light). Suppose now we encounter a tiny dust grain whose mass is a millionth of a gram. This dust grain had better be absorbed by a dust-shield, since otherwise it will certainly cause severe damage to the equipment (and possibly us). If it gets absorbed it transfers its whole energy and momentum to the dust-shield. Its kinetic energy is $m_0 c^2 (\gamma - 1)$, that is four times its rest energy; cf. (3.17). This makes four-hundred million joules ($4 \cdot 10^8$ J), which corresponds to the kinetic energy of a Rolls-Royce luxury limousine, of approximately three tonnes weight, at twice supersonic speed! How should one protect oneself from such a bombardment?

Even more questionable is the realization of an engine that can provide the required acceleration over such an extended period. Note that $\gamma - 1$ is the kinetic energy in units of the rest energy;

cf. (3.17). On reaching $\gamma = 5$ the engine must have converted 80% (i.e. 4/5) of the original total mass into kinetic energy, which seems technically totally out of reach. As we have seen in Sect. 4.2, nuclear reactions turn mass into energy at an efficiency of at most a few per cent. Hence the engine can certainly not be based on nuclear energy. In principle, up to 100% efficiency can be reached by letting matter annihilate the same amount of antimatter. If the hard γ-rays that are produced during annihilation could then be reflected into one direction, the recoil of the γ-rays could then provide the necessary thrust. But no 'mirror' is known to exist that can reflect γ-rays. Moreover, how should one ever be able to store large amounts of matter and antimatter in the same spaceship, given that they may nowhere be in contact except for the burning chamber, for otherwise they would immediately annihilate? The answers seem to be known only to science fiction authors. Seriously, there is not the slightest evidence that we will ever make it even to our nearest neighbouring star, Proxima Centauri, let alone cruise the Universe.

4.6 Outlook on General Relativity

As emphasized many times, the physical validity of SR is limited to all those phenomena in which gravity can be neglected. If gravity gets involved, we have to resort to the theory of General Relativity. Being a generalization of SR, it is possible to understand, at least qualitatively, some of the characteristic features that distinguish General Relativity from Newtonian gravity within the framework of SR through a careful admixture of some heuristic ideas. This we wish to do in this section.

One of the most spectacular predictions of General Relativity is that of gravitational collapse. Roughly speaking it says that sufficiently compressed masses will inevitably collapse to form black holes. Whatever material they may be made of, no internal pressure can ever stop this collapse. The exact value of the critical radius, below which the collapse of a mass M necessarily sets in, depends a little on the type of matter, but it is always of the order of the gravitational radius $R_G = GM/c^2$ that we already encountered in Sect. 4.4. The gravitational radius of the Earth is 4.4 millimetres, as already mentioned; that for the Sun is 1.5 kilometres.

In Newtonian physics we can always balance the inwardly directed gravitational pull of a star's atmosphere by letting the star contract sufficiently, though, not endlessly. The point being that, upon contraction, the outwardly directed internal pressure grows faster than the gravitational pull, thereby eventually leading to a cancellation. This is different in General Relativity, which predicts a stronger than Newtonian increase of the gravitational pull in the contraction phase. To some extent, the reason for this is of special relativistic origin. More precisely, it has to do with the prediction of SR that a body put under material stresses has a larger inertial mass than the same amount of matter without stresses. This will be shown below. Given that, we only need to add one basic principle of General Relativity, according to which inertial and gravitational masses are universally proportional to each other ('universally' meaning that the constant of proportionality is the same for all types of matter; it can hence be set to 1 by an appropriate choice of units). It follows that material stresses add to the gravitational field of the body.

This gives rise to the following collapse scenario. Consider a cloud of initially almost pressureless gas that starts to contract under its own gravitational pull. The decrease in volume pushes up the gas pressure. Pressure is a special form of material stress that adds to the gravitational field of the material and hence to the gravitational pull. This additional pull causes further contraction of the star, beyond the point where a Newtonian equilibrium is reached. But this further contraction causes still higher pressures, and therefore, in turn, still higher gravitational pull. This process now iterates. Under certain circumstances this might have the effect that the gravitational pull grows faster than the internal gas pressure. Then no equilibrium exists and the star inevitably collapses. The stationary end product of such a collapse will be a black hole.

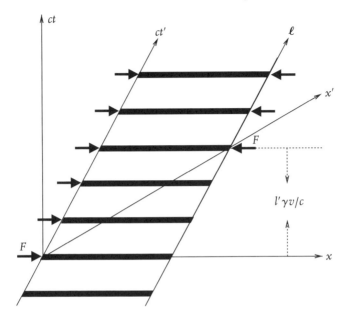

Fig. 4.5 Moving rod under tension.

It remains to understand that SR indeed predicts that material stresses add to the inertial mass of a body. This we do by means of a special example depicted in **Fig. 4.5**. Consider a homogeneous cylindrical rod of length l' and cross-section q' that rests along the x'-axis in system K'. As always, K' moves relative to K with velocity v in the x direction, such that both systems coincide at $t = t' = 0$. The world line of the rod's trailing end is just the ct'-axis, that of its leading end is denoted by ℓ. Now assume some mechanism by means of which two equal and opposite (inwardly) directed forces are simultaneously (in K') applied to the two ends of the rod. Let the forces start to act at $t' = 0$. It is obvious that these forces will not set the rod into motion relative to K', since in K' they start to act simultaneously with equal strength but opposite directions. **Figure 4.5** now shows the same process relative to K. The acting forces are depicted by the arrows. The crucial point is that now the forces do not start to act simultaneously on both ends with respect to time t. Rather, the pushing force starts to act on the trailing end at $t = 0$, whereas the counteracting force starts to act on the leading end at the later time $t = l'\gamma v/c^2$. This immediately follows from the Lorentz transformation (3.3) by setting $x' = l'$ and $t' = 0$. In the meantime the observer in K reckons the rod to be one-sidedly pushed in the direction of motion with a force of strength $F = F'$. This equality means that longitudinal forces are measured with the same strengths in K and K'. This follows from the first transformation rule for forces, (5.39), to be derived later, by specializing it to $\vec{u}' = 0$ since the rod rests in K'. Hence, reckoned from K, there is a transfer of longitudinal momentum of $\Delta p = F'l'\gamma v/c^2$. But this additional momentum does not lead to any increase in the velocity of the rod. Hence its inertial rest-mass must have increased by $\Delta m_0 = F'l'/c^2$. Since, in K', $q'l'$ is the volume of the rod and $F'/q' = p'$ is the pressure (force per unit cross-section), one may also say that the rest-mass density increased by p'/c^2.

This is precisely the pressure term that appears in addition to the ordinary mass term as a source for the gravitational field in General Relativity. In particular, it appears as a contribution of the radial stresses in the so-called Oppenheimer–Volkov equation, which governs the equilibrium configurations of spherically symmetric stars. Moreover, this term also plays a crucial rôle in cosmology, where thrice it (the sum of the stresses over all spatial directions) is added to the mass density in the so-called Friedmann equations that govern the expansion of the Universe on the largest scales. Here it is not the pressure of ordinary matter, which is far too small, that takes dynamical influence, but there is the so-called 'dark energy', sometimes also referred to as the 'cosmological constant' Λ, which has the meaning of a constant (in space and time) positive-energy density, ρ_Λ, and associated to it a very large *negative* pressure, $p_\Lambda = -\rho_\Lambda$. The relevant combination for the Friedmann equations is $\rho_\Lambda + 3p_\Lambda$, which equals $-2\rho_\Lambda$. Recent cosmological measurements of various kinds led to the conclusion that approximately

70% of all gravitating energy is in the form of a cosmological constant, whereas only one-sixth of the remaining 30% (i.e. 5% in total) is localized in 'normal matter', i.e. atoms that make up the world around us. The question of what the remaining 5/6 of the non-cosmological-constant matter may be is known as the 'dark-matter problem', whose answer is still open to speculation. In any case, it now seems reasonably well established that the current state of the Universe is one of accelerated expansion, driven by the negative pressure p_Λ. This leaves unanswered the question of what the origin of a cosmological constant of that size might be (known as the 'dark-energy problem'), which is currently felt to be one of the most challenging problems in theoretical physics. Compare the remarks made at the end of Sect. 4.3.

·5·
Closer encounters with special topics

5.1 Ole Rømer's measurement of the velocity of light

The first reliable measurement of the speed of light which, in particular, proved it to be *finite*, was performed by the Danish astronomer Ole Rømer in the years 1672–76. The basic idea is reported to go back to Giovanni Domenico Cassini (1625–1712) and will be described in this section.

In 1610 Galileo Galilei discovered the four largest moons of the planet Jupiter, which he called the 'Medicean Stars'. Beginning with the innermost, they are now called Io, Europa, Ganymede, and Callisto. They can quite easily be seen using a good pair of field glasses. Today 59 additional Jovian moons are known to exist, all of which are considerably smaller than the four Galilean ones. In fact, 48 of them have a diameter of less than 10 kilometers. The Galilean moons have nearly circular orbits (eccentricities of a few times 10^{-3}) whose inclination against Jupiter's equator is also very small. This is particularly true for Io, whose orbital inclination is a tiny 0.04°. (Jupiter's orbital plane has an inclination against the ecliptic—the orbital plane of the Earth—of 3.1°.) The periods of the four Galilean moons lie between 1.769 days for Io and 16.689 days for Callisto. More precisely, these are the so-called sidereal periods, after which Io and Jupiter bear the same geometric relation to the fixed star.

Rømer measured the orbital periods of Io and discovered an apparent regular variation whose period was one year. He (and Cassini) correctly interpreted this as an effect caused by the finiteness of the speed of light. In phases where the Earth approached Jupiter, Io's periods seems to be smaller, and larger in period where

Measurement of the velocity of light

the separation between Jupiter and the Earth increased. In order to make such a statement concerning Io's periods, Rømer had to come up with some operational means to decide, from the standpoint of the Earth, when Io had gone 'once around' Jupiter. In other words, he had to 'mark' a particular point on Io's orbit. Rømer's solution was to consider that orbital point where Io enters the shadow cast by Jupiter on its far side, as seen from the Sun. The corresponding eclipse of Io is clearly an event that can quite easily be located in time by an observer on Earth. But note that this does not quite coincide with the sidereal period of Io. The period measured by Rømer is that after which Io, Jupiter, and the Sun (rather than the fixed stars) bear the same geometric relation. It is called the synodic period and differs from the sidereal period due to Jupiter's orbital motion around the Sun. Since the latter is in the same direction as Io's motion around Jupiter, the synodic period exceeds the sidereal period by a small amount. Roughly speaking, after one sidereal period of Io, Jupiter has advanced by a small amount along its orbit, which Io needs to catch up with in order to come into the same relative geometric configuration with respect to the Sun and Jupiter. More precisely, the relative excess of Io's synodic over its sidereal period is given by the ratio of Io's to Jupiter's sidereal periods. This fraction is approximately $1.7/12 \cdot 365 \approx 4 \cdot 10^{-4}$, so that the synodic period of Io exceeds its sidereal one by just about a minute.

For a more detailed discussion we consider **Fig. 5.1**, where we represent the three orbits—Earth around the Sun, Jupiter around the Sun, and Io around Jupiter—as if they were in a common plane. This approximation is allowed in the present context. As an example we depicted a phase in which the Earth approaches Jupiter. Let E_0 be the position of the Earth and T_0 the time reading on Earth where the first Io eclipse is registered. At this time the distance between the Earth and Jupiter is l_0 (and with sufficient approximation also between the Earth and Io). Now one waits for n more eclipses of Io to occur. Let E_n and T_n be the Earth's position and time at which this nth further eclipse is seen on Earth. At this time the distance from the Earth to Jupiter is l_n. Hence, in the meantime, the distance has changed by $l_n - l_0$, mainly due to the orbital motion of the Earth (and

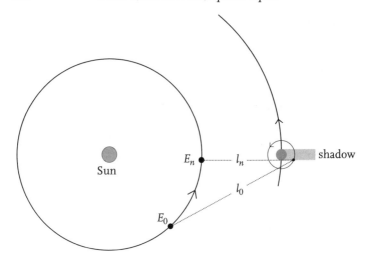

Fig. 5.1 Rømer's experiment to measure the speed of light.

a little due to Jupiter's orbital motion). This difference is negative since l_n is smaller than l_0. If τ denotes the synodic period of Io, we have

$$T_n = T_0 + n\tau + \frac{l_n - l_0}{c}. \tag{5.1}$$

The third term on the right-hand side takes into account the change in the travel time of light due to the changing distance between the Earth and Jupiter. Formula (5.1) is now read as an equation in the two unknowns: τ, the synodic period of Io, and c, the velocity of light. The idea is then to apply (5.1) to two different measurements, so as to obtain two equations for two unknowns. In the first measurement one counts the total number, N, of Io eclipses during the time the Earth goes once around the Sun and returns to the same configuration with respect to the Sun and Jupiter. This is called the synodic period of Jupiter and corresponds to 399 days. It is longer than the sidereal year of the Earth because of Jupiter's orbital motion (now the Earth has to catch up with Jupiter). In this

Measurement of the velocity of light

case we have $l_0 = l_N$ so that (5.1) gives an expression for τ:

$$\tau = \frac{399 \text{ days}}{N}. \tag{5.2}$$

In the second measurement one counts the number, N', of Io eclipses during half a synodic period of Jupiter, beginning, say, with the position E_0 where the Earth has the largest distance to Jupiter. (Here we cheat a little, because Jupiter is hardly visible if the line of sight to Jupiter runs close to the Sun. For the sake of an easy argument we shall neglect this point.) After 399/2 days the Earth is closer to Jupiter by an amount that corresponds to the Earth's orbital diameter. By definition, the (mean) orbital radius of the Earth is a so-called astronomical unit, AU, which is approximately given by 150 million kilometres. Hence $l_0 - l_{N'}$ is given by 300 million kilometres. Applying (5.1) to this second case leads to an equation for the difference of N' synodic periods of Io to half a synodic period of Jupiter:

$$N'\tau - \frac{399}{2} \text{ days} = \frac{2 \text{ AU}}{c}. \tag{5.3}$$

Inserting the value for τ from the first measurement allows us to compute the left side of this equation. Modern measurements give approximately 17 minutes. Hence one can conclude that light needs 17 minutes, or 1020 seconds, to travel a distance of two astronomical units. In other words, in one second light travels approximately the thousandth part of 2 AU = 300 million kilometres, that is, 300 thousand kilometres. Originally Rømer obtained only 3/4 of that value because of observational errors in the timing of the eclipses and also because he did not know the exact value for the astronomical unit (which was measured by Cassini in 1672 using the parallax of Mars).

The main result was not so much the precise value for c, but rather that from now on one knew with certainty that the speed of light is finite. In those days no refined methods existed to measure very short time intervals. Hence one had to resort to astronomical methods in order to provide measurable travel times of light. It was not before 1849, 170 years after Rømer's measurements, that Fizeau used terrestrial methods to measure c. He obtained a value that was just 5% above the exact one cited in (1.1).

5.2 The independence of the velocity of light from the state of motion of the source

In an ether theory, light corresponds to elastic waves in a real medium, the ether, just like sound corresponds to waves in air or water. If the classical ether theory is correct, light should always have the same velocity relative to the ether (neglecting dispersion phenomena for the moment), independently of the state of motion of the source. As an alternative to the ether theory, it was suggested that the emission of light corresponds somehow more to a ballistic process, in which light should always have the same velocity relative to the emitting agent. At first sight this idea seems to fit more into a particle picture of light, which was already overturned in the 19th century. But there was indeed a way to also incorporate the ballistic idea into a wave theory. Such a theory was developed by the Swiss physicist Walter Ritz in 1909 (five years after the publication of SR!), who essentially took Maxwell's equations (in integral form) and made a few rather bold formal changes that led to the desired (from his point of view) changes in wave propagation. His motivation, shared by many of his contemporary physicists, was to overcome the special relativistic notion of time and to return to the classical notion of absolute time.

According to Ritz's theory, as well as all other so-called 'emission theories', the velocity of light relative to a spatially fixed observer depends on the state of motion of the emitting agent. This prediction can be tested on astronomical objects, as was first pointed out in 1913 by the Dutch astronomer Willem de Sitter (1872–1934) [20]. Let us illustrate his idea by means of **Fig. 5.2**. Consider a double-star system in which two stars orbit their common centre-of-mass. To keep the discussion simple, we assume one companion to be significantly heavier than the other one, so that its location

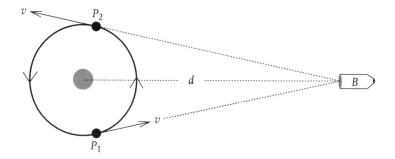

Fig. 5.2 De Sitter's experiment.

practically coincides with the centre of mass. Accordingly, **Fig. 5.2** shows one star resting in the centre, and another star that revolves around it counter-clockwise on a circular orbit with velocity v. An observer B is positioned at a distance d in the orbital plane of the system, observing the double-star system in his telescope. The distance d is assumed much larger than the orbital diameter of the double-star system. If the orbiting star is at position P_1 it directly heads toward the observer. Emission theories predict that light sent from P_1 approaches the observer at a higher speed than c, in contrast to SR. We denote this speed by $c + kv$, where k is a parameter that interpolates between SR ($k = 0$) and Ritz's theory ($k = 1$). k may be thought of as parametrizing the degree to which the velocity of the source adds to the velocity of the emitted light. The time light travels from P_1 to the observer B is given by $T_1 = d/(c + kv)$. Similarly, at P_2 the star directly moves away from the observer. Light emitted from P_2 approaches the observer at speed $c - kv$. Its travel time is now given by $T_2 = d/(c - kv)$. Since we assumed d to be much larger than the orbital diameter, P_1 and P_2 are almost diametrically opposite orbital points. (This does not come out very well in our drawing, since we greatly exaggerated the orbital diameter in relation to d.) Hence, assuming a strictly circular orbit, the star takes the same time from P_1 to P_2 as it takes from P_2 to P_1, namely half its period, $T/2$. (We denote the period by T.) But this is not what is *seen* by the observer B. For suppose at time $t = 0$

the star is at P_1. Then B sees that event at the later time $t = T_1$. At time $t = T/2$ the star is at P_2. This event is seen by B at time $t = T/2 + T_2$. Hence the travel time from P_1 to P_2, as seen and measured by B, is given by

$$T_{12} = \frac{T}{2} + T_2 - T_1 \approx \frac{T}{2} + k\frac{2vd}{c^2}, \tag{5.4}$$

which is longer than the 'true' semi-period $T/2$. Here $\approx$ indicates an approximation where second and higher powers in v/c are neglected (we are just interested in the leading order). Similarly, the *observed* travel time from P_2 to P_1 is

$$T_{21} \approx \frac{T}{2} - k\frac{2vd}{c^2}, \tag{5.5}$$

which is shorter than the 'true' semi-period $T/2$. According to these results the observer should see a fairly irregular motion, in which the star takes more time to traverse the semi-circle P_1P_2 pointing toward the observer, and less time for the complementary semi-circle P_2P_1. It might even happen that light sent from P_1 in the nth cycle catches up the light sent before from P_2 in the $(n-1)$st cycle. Indeed, spectroscopic observations allow us to determine v via the Doppler shift of spectral lines. d can be measured independently. In many cases the correction term $2vd/c^2$ turns out to be of the same order as, or even bigger than, the semi-period $T/2$. But even for such systems no anomalous orbital motion was ever detected.

From his observational material de Sitter already concluded that the parameter k must be smaller than $1/2000$, which would clearly refute Ritz's theory. But it was also argued, quite correctly, that his observations do not warrant this conclusion, the reason being as follows: de Sitter exclusively used light in the optical part of the electromagnetic spectrum. The interaction of interstellar matter with electromagnetic waves in that frequency range is sufficiently strong for the light to have undergone several (in the mean) absorption and re-emission processes on its way from the star to the observer. But then, according to the emission theory, the additional velocity of the emitted light would not be that of the orbiting star but rather the mean velocity of the interstellar medium. But the latter should be constant relative to the Earth, so that none of the effects described above should occur, in *agreement* with actual observations.

For this reason de Sitter's experiments were repeated in 1977, this time using X-ray binary systems. These are double-star systems in which one companion is a 'pulsar', that is a star that periodically sends electromagnetic pulses in the X-ray regime, which lies at much higher frequencies than the optical regime. The point being that the interaction between interstellar matter and X-rays is so much smaller that it gives rise to almost no absorption process (on average) on the way to the observer. Now de Sitter's original argument does indeed apply. But this time, too, no anomalous orbital motion was seen. These results now put a very stringent upper bound on the value of k [21]:

$$|k| < 2 \cdot 10^{-9}. \tag{5.6}$$

There are also terrestrial experiments involving fast moving particles emitting light. But so far these could not improve on the bound set by (5.6).

5.3 Do superluminal velocities exist?

If posed in that generality, the question has to be answered by a clear 'yes'. Special Relativity puts the upper limit c only for propagation speeds of particular processes. These include the motions of material bodies and, more generally, all processes which, in principle, can be used to transmit signals. To be sure, for this statement to make unambiguous physical sense one would have to give a clear-cut definition of 'signal', which we will not attempt here. In any case, is should be clear from the discussion in Sect. 3.5 that superluminal signal propagation will lead to severe difficulties with the causality relations imposed by SR, whatever the precise definition of signal may be.

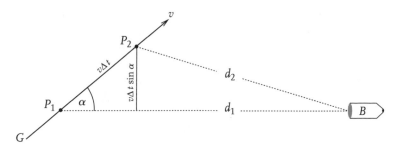

Fig. 5.3 The origin of apparent superluminal velocities.

Let us first consider an example taken from modern astronomy. It serves to illustrate the fact that it is easy to create the visual impression of superluminal propagations, even though neither light nor any of the bodies involved propagate faster than c. In **Fig. 5.3** we sketch a situation in which some luminous material object propagates along a straight line G with velocity v, thereby approaching an observer B at some acute angle α. Let P_1 be the position of the object at time t_1. At this moment its distance to the observer is given by d_1. Light emitted at P_1 reaches the observer at the time $t_1^{(B)} = t_1 + d_1/c$, according to SR (independent of the velocity of the source). Consider a small time interval, Δt, in which the body moves from P_1 to P_2. Now the distance to the observer is d_2. Light emitted at P_2 reaches the observer at time $t_2^{(B)} = t_1 + \Delta t + d_2/c$. The time interval that the observer measures between these two light signals is therefore given by $\Delta t^{(B)} = t_2^{(B)} - t_1^{(B)} = \Delta t - (d_1 - d_2)/c$. Now, the crucial point is that d_2 is smaller than d_1. Explicitly we have, up to higher than linear terms in Δt:

$$d_2 = d_1 - v\Delta t \cos\alpha. \tag{5.7}$$

This means that the signal sent at P_2 has a shorter travel time to the observer than the signal sent at P_1. The observer thus sees the object propagating from P_1 to P_2 within a time interval $\Delta t^{(B)}$ that is *shorter* than Δt. For small Δt we have in the linear approximation, using (5.7),

$$\Delta t^{(B)} = \Delta t - (d_1 - d_2)/c = \Delta t(1 - \beta \cos\alpha). \qquad (5.8)$$

Here we again wrote β for v/c. During this time interval the observer sees the object passing a distance $D = v\Delta t \sin\alpha$ perpendicular to his line of sight. This corresponds to an apparent transverse visual velocity of

$$v_B = \frac{D}{\Delta t^{(B)}} = c \cdot \frac{\beta \sin\alpha}{1 - \beta \cos\alpha}. \qquad (5.9)$$

It is easy to see that the factor multiplying c on the right-hand side is unbounded from above. For fixed β and variable α it assumes its maximum at $\cos\alpha = \beta$. The corresponding maximal value is $v_B^{\max} = \gamma v$, which diverges if v tends to c.

Today many astronomical examples of this effect are known. A particularly impressive one is given by the galaxy M87 [22], which is located in the Virgo cluster at a distance of approximately 60 million light years from us. There are two jets of highly accelerated gas emerging from the central region of this galaxy, pointing in opposite directions perpendicular to the galactic plane. The visual velocity v_B of the jet stream is six times c. They are possibly generated by a supermassive black hole that is conjectured to reside in the centre of M87. The actual velocity, v, of the gas jet is estimated to be at most 0.98 c. For more details, see e.g. [23].

We now wish to turn to some more fundamental aspects in connection with superluminal speeds, whose non-observance has in recent years led to some turmoil in the daily press. Let us start with the following general statement, that it is not an entirely obvious matter to assign a single characteristic velocity to an extended physical entity. If this entity is a body, we may try to define it through the velocity of its centre of mass. But let it be said in passing that 'centre of mass' is not an entirely unproblematic notion in relativistic kinematics. This general remark is particularly true if the entity is a wave. There are many possibilities to assign a velocity to a wave, none of which is obviously distinguished. Mathematically, a general wave is obtained by superposing pure sinusoidal waves of different frequencies and infinite extent. This is called Fourier composition, after the French mathematician Jean Baptiste Joseph Fourier (1768–1830). Each Fourier component, also called a partial wave, has a fixed frequency and wavelength. Their phases propagate with a well defined velocity, called the *phase velocity*. For light waves it is given by c/n, where n is the index of refraction of the medium in which the light propagates. However, n generally also depends on the frequency of the light wave, a phenomenon known as dispersion (cf. Sect. 2.4.2). More precisely one speaks of normal/anomalous dispersion if n grows/decays with increasing frequency. It is clear that a single sinusoidal wave of infinite extent and no further structure cannot be used for transmitting signals. Hence SR does not rule out phase velocities greater than c. And, indeed, situations where $n < 1$ frequently occur.

By adding purely sinusoidal waves one can modulate structures (humps, wave packets) which can be mathematically assigned a centre. The centre moves with the so-called *group velocity*. In a restricted sense, such wave packets can be used for the transmission of signals. The restriction is a result of dispersion, that is, different phase velocities for the partial waves, which quite literally leads to 'dispersions of the wave packets'. This is where the notion of dispersion receives its name from. Signal transmission works as long as the structure of the wave packet is sufficiently stable against dispersion, at least for the time of transmission. On the other hand, signal transmission is clearly out of the question if dispersion lets the wave packet decay within a time in which its centre has just moved a single packet width. This is a very important though somewhat subtle point. Purely formally we can always define a centre of the wave packet, however spread it might be, and calculate its velocity. But only in regimes of sufficiently small dispersion can this velocity be identified with a physical signal velocity!

Do superluminal velocities exist?

This has led to the confusion alluded to above, since the group velocity can, in fact, become larger than c, but only in regimes where dispersion is too strong for physical signal transmission. Disregarding this point has sometimes led to claims that *signals* (and even whole Mozart symphonies) had been transmitted faster than light, and that hence SR had been falsified. But this is definitely not true.

An obvious transmission of a signal is achieved by sending a wave packet whose amplitude is non-zero only for a finite time interval. Here we might think, for example, of the Morse alphabet, where single letters are separated by finite time intervals of absolute silence. ('Absolute silence' may also be understood as a purely stochastically fluctuating background, as e.g. in Quantum Electrodynamics.) How fast can a single letter of that alphabet be transmitted? A natural velocity to be assigned to such a letter is the *front velocity*. This is the velocity by which the foremost (in the direction of propagation) non-zero amplitude propagates. The front velocity is the most natural candidate to be identified with a physical 'signal velocity'.

Finally there is the *energy velocity* by which energy is transmitted in the wave field. It is not necessarily identical with any of the aforementioned velocities.

Phase and group velocities exceeding c do not constitute a contradiction to SR. This would be different for front or energy velocities. Any of them exceeding c would pose severe problems for SR. But so far there are absolutely no signs to this effect, neither experimental nor theoretical. **Figure 5.4** shows the typical behaviour of all these velocities in the 'dangerous' frequency region where anomalous dispersion occurs. We plotted the ratio c/u as a function of the frequency, where u is any of the four velocities discussed here. For phase and group velocities this quotient may become less than one, but always stays above the line $c/u = 1$ if u is the front or energy velocity. For more information on the topic of this section, see [24].

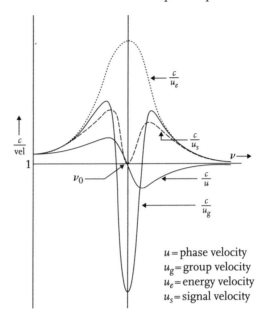

Fig. 5.4 Typical behaviour of different velocity types around the region of anomalous dispersion (critical frequency v_0).

5.4 The Kennedy–Thorndike experiment

The negative result of the Michelson–Morley experiment shows the isotropy of the speed of light, that is, its independence from the orientation of the interferometer. To be precise, it shows the independence for the mean velocity, averaged over both directions in the arm of the interferometer. This clearly speaks against the existence of an ether, but does not prove it. For this one would have to prove that no measurements whatsoever of the speed of light could disclose a somehow preferred frame of reference. For notational simplicity we continue to call this hypothetically preferred frame the 'ether system', even though we do not need to think of a material ether as being the actual cause of this preference. To say it once more, the Michelson–Morley experiment shows that the speed of light measured by some observer does not depend on the direction of the observer's velocity relative to the ether system. But it does *not* show that it is independent of the magnitude as well. This is achieved by the experiment performed by Roy Kennedy and Edward Thorndike in 1932 [25], which we will now discuss.

The basic equipment is again an interferometer of the type used by Michelson and Morley; cf. Sect. 2.4.3. But this time, for reasons that will become clear soon, the lengths of the two arms are chosen to differ significantly. (In the original Michelson–Morley experiment they were approximately equal.) Also, in the Kennedy–Thorndike experiment, one is not interested whether shifts of interference fringes occur during rotations of the apparatus. Rather, one looks out for such shifts in the course of a much longer time, during which the apparatus has changed the magnitude of its velocity relative to the ether system appreciably. To see how such a change comes about, we recall that the velocity $\vec{v}$ of the apparatus relative to the ether system can be decomposed into three components:

1. The velocity $\vec{v}_R$ of the apparatus relative to the centre of the Earth, due to its daily rotation.
2. The velocity $\vec{v}_E$ of the Earth's centre relative to the Sun, due to its annual orbital motion.
3. The velocity $\vec{v}_S$ of the Sun relative to the ether system.

Hence we have

$$\vec{v} = \vec{v}_R + \vec{v}_E + \vec{v}_S. \tag{5.10}$$

Kennedy and Thorndike assumed the magnitude of $\vec{v}_S$ to be significantly larger than the magnitudes of the other components, comparable to the then largest known relative velocities of astronomical objects of a few hundred kilometres per second. Today we know that our solar system orbits the galactic centre at a speed of approximately 220 kilometres per second. Much higher relative velocities between galaxies have also been observed. What, then, would be a plausible value for the magnitude of $\vec{v}_S$?

Here modern cosmology gives a strong hint. The whole Universe is filled with a faint background of radio waves, peaked about a wavelength of 2 millimetres. It is called the *cosmic microwave background radiation*. Such a space-filling radiation defines a preferred system of reference, namely that in which the radiation looks isotropic. Hence one may define one's velocity 'relative to the radiation'. Our velocity has, in fact, been measured by the satellite COBE (cosmic background explorer) in the early 1990s, which took precise data of the power spectrum of this radiation in various directions. The result was that our solar system moves relative to the radiation with a velocity of 380 kilometres per second. (This velocity is directed almost oppositely to our velocity relative to the galactic centre, which means that the latter moves relative to the radiation by 600 kilometres per second.) In the rest system of the radiation the spectrum is very closely (up to one part in 10^5) that of a black body of temperature 2.73 degrees Kelvin. Modern big-bang cosmology relates the origin of this radiation to the evolutionary phase where free electrons and nuclei combined into stable atoms under the emission of light. From that time on—approximately 300 000 years after the big bang—the interaction of light with matter is strongly suppressed and the Universe becomes 'transparent'. These fundamental considerations make the rest system of the microwave background radiation a strong candidate for the identification of a hypothetical ether system, though, strictly speaking, there is no logical necessity in this. Accordingly we take the magnitude of $\vec{v}_S$ to be 380 kilometres per second.

In the course of 12 hours $\vec{v}_R$ changes to $-\vec{v}_R$, whereas the other components of $\vec{v}$ stay nearly constant. The daily variation of the squared velocity,

$$v^2 = (\vec{v}_R + \vec{v}_E + \vec{v}_S)^2, \tag{5.11}$$

is then given by

$$\Delta v^2 = 4\vec{v}_R \cdot (\vec{v}_E + \vec{v}_S). \tag{5.12}$$

In the same fashion $\vec{v}_E$ changes to $-\vec{v}_E$ within 6 months, whereas $\vec{v}_S$ stays constant and the precise timing can be chosen such that $\vec{v}_R$ returns to its initial value (though this does not really matter, due to the much larger modulus of $\vec{v}_S$). This produces an annual

The Kennedy–Thorndike experiment 125

variation of the squared velocity, given by

$$\Delta v^2 = 4\vec{v}_E \cdot (\vec{v}_R + \vec{v}_S). \tag{5.13}$$

Recall that the (expected) modulus of $\vec{v}_S$ (380 km/s) is considerably larger than the moduli of $\vec{v}_E$ (30 km/s) and $\vec{v}_R$ (0.46 km/s at the equator). Hence expressions (5.12) and (5.13) are essentially dominated by the projections of $\vec{v}_S$ into the equatorial plane of the Earth and the ecliptic, respectively. The Kennedy–Thorndike experiment puts restrictions on the magnitude of these projections.

According to (2.16) and (2.13), (2.15), the difference in the number of phases along both interferometer arms is given by

$$N = \nu\gamma \frac{2(\gamma l_1 - l_2)}{c}. \tag{5.14}$$

Now we invoke the general deformation hypothesis (cf. Sect. 2.4.4), which explained the negative result of the Michelson–Morley experiment. Accordingly, we replace the lengths l_1 and l_2 in (5.14) by Al_1^0 and Bl_2^0, respectively, which merely express the general deformation hypothesis (2.20). Then we invoke the relation (2.22), which expresses the result of the Michelson–Morley experiment. The latter we use in the form $A = B/\gamma$ to eliminate A in favour of B. This leads to

$$N = \nu\gamma B \frac{2(l_1^0 - l_2^0)}{c}. \tag{5.15}$$

Here the factor γ on the right-hand side is a function of v^2, as written down explicitly in (2.14). A daily or annual variation of this squared velocity as in (5.12) or (5.13) should therefore also result in a corresponding variation of N according to (5.15), and hence to an observable shift in the interference pattern. Since this shift is proportional to the difference $l_1^0 - l_2^0$, the arm lengths had to be chosen as different as possible. However, the Kennedy–Thorndike experiment did not reveal any significant effect. More precisely, the data for the daily variation led to an average modulus of 24 kilometres per second for the projection of $\vec{v}_S$ into the equatorial plane of the Earth. The data for the annual variation gave a corresponding value of 15 kilometres per second for the ecliptic projection of $\vec{v}_S$. These values lie more than one order of magnitude below those expected from an ether drift. Moreover, taking into account all possible errors, this result is shown to be compatible with a null result. This is supported by the fact that the determined average directions for $\vec{v}_S$'s equatorial and ecliptic projections enclose an angle of 123 degrees, which certainly does not speak for a systematic ether drift. Admittedly, all this sounds like a rather imprecise measurement. But one should not forget that the experimental demands were quite exceptional, and so were the achievements to meet them. The main challenge was to keep the delicate experimental environment extremely stable over many days and even months! For example, the experimenters succeeded in restricting temperature variations to the order of millidegrees during the whole period of data taking, against the natural daily and long term variations.

From their experiment, Kennedy and Thorndike concluded that N (cf. 5.15) would not depend on v^2 at all. Moreover, they assumed $B = 1$, even though they knew (as stated explicitly in their paper) that this does not follow from the experiment of Michelson and Morley. This assumption led them to conclude from (5.15) that the combination $\nu\gamma$ must be a v^2-independent quantity, which we call ν'. Hence they obtained

$$\nu = \nu'/\gamma = \nu' \cdot \sqrt{1 - \frac{v^2}{c^2}}. \quad (5.16)$$

Recall that the frequency ν denotes the number of oscillation periods within one unit of time measured with clocks in the ether system K. But the light source rests in system K' and its frequency is defined with respect to the clocks that rest in K'. (Kennedy and Thorndike used light where $\lambda' = 5461$ Å, corresponding to a green line of mercury.) But then (5.16) just corresponds to the statement of time dilation, where ν' is the fixed frequency of the source in K'. Hence the null result of the Kennedy–Thorndike experiment is indeed implied by time dilation and $B = 1$. However, Kennedy and Thorndike concluded the logical converse of this, namely that their experiment implied time dilation. (This attitude is already expressed by the title of their paper, reading 'Experimental Establishment of the Relativity of Time'.) But this would be logically correct only if $B = 1$ had already been experimentally established, rather than merely assumed, and this was not the case at the time (1932) Kennedy and Thorndike presented their results. For this reason we merely deduce from their experiment that (5.15) leads to the more general (i.e. logically weaker) conclusion

$$\nu = \nu'/\gamma \cdot B, \qquad (5.17)$$

where ν' is velocity independent. In the next section we will discuss the experiment of Ives and Stilwell, performed six years after that of Kennedy and Thorndike, which independently, and now truly, led to (5.16). Using this, we may then employ the result (5.17) of Kennedy and Thorndike to conclude $B = 1$.

5.5 The Ives–Stilwell experiment

In 1938 Herbert Ives and G. R. Stilwell performed an experiment [26] which was designed to test the relativistic Doppler effect, as given by formula (3.10). More precisely, they aimed to verify the occurrence of the factor γ in the denominator. This factor is a consequence of time dilation and therefore a genuine effect of SR, whereas the other dependencies expressed by this formula are just the same as in 'pre-relativistic' physics (cf. the discussion in Sec. 3.6.2). Already in 1907 Einstein [12] suggested testing the occurrence of γ through the transverse Doppler effect (3.11). But this idea had to be given up soon for the following reason. The transverse Doppler effect is of second order in $\beta = v/c$. This means that in a perturbation expansion in β, the leading term is quadratic. In contrast, the longitudinal Doppler effect is of first order in β and hence generically much larger, at least as long as the velocities involved are not very close to c. According to (3.10) any small deviation δ_{90} of the observing angle α from 90° leads to admixtures of the longitudinal Doppler effect which will immediately 'swamp out' the transverse effect. Note that these admixtures are in leading order also proportional to the deviation angle δ_{90} (see below). But such small deviations are hardly avoidable. For example, even the finest pencil of fast moving molecules has a small opening angle due to residual transverse velocities. Hence some of the molecular rays will not exactly be seen at 90°. These suffice to spoil the direct experimental observation of the transverse Doppler effect.

This problem was overcome by Ives and Stilwell. Instead of trying to observe at right angles in order to isolate the γ dependence, they observed at 0° *and* 180°. To see why this solves the problem, we first have to rewrite (3.10) in terms of wavelengths, since this is what Ives and Stilwell actually observed (not frequencies). This is easily done using $\nu = c/\lambda$ and $\nu' = c/\lambda'$:

$$\lambda = \lambda'\gamma(1 + \beta \cos \alpha). \tag{5.18}$$

Now set α equal 0° and 180° and call the corresponding wavelengths λ_0 and λ_{180} respectively. We get

$$\lambda_0 = \lambda'\gamma(1 + \beta) \quad \text{and} \quad \lambda_{180} = \lambda'\gamma(1 - \beta). \tag{5.19}$$

It is now easy to see how to eliminate the linear term in β and isolate the γ dependence. One simply takes the mean:

$$\tfrac{1}{2}(\lambda_0 + \lambda_{180}) = \gamma\lambda'. \tag{5.20}$$

Without time dilation the mean wavelength would be just that in the rest system of the light source. But taking into account time dilation from SR enhances the mean value by a factor of γ. This is what Ives and Stilwell measured.

To be sure, here too one needs to take into account possible deviations δ_0 and δ_{180} from the exact values $\alpha = 0°$ and $\alpha = 180°$, respectively. These give rise to additive correction terms on the right-hand side of (5.20), of which the leading term reads:

$$-(\delta_0^2 - \delta_{180}^2)\frac{\beta\gamma\lambda'}{2}. \tag{5.21}$$

Let us compare this to the above-mentioned correction for an observation at $\alpha = 90°$. At exactly 90° (5.18) leads to $\lambda = \gamma\lambda'$.

At $\alpha = 90° + \delta_{90}$ the right-hand side receives a correction of

$$-\delta_{90}\beta\gamma\lambda'. \tag{5.22}$$

Both corrections are *linear* in β. This is what makes them potentially fatal. However, (5.21) is quadratic in the small deviation parameters whereas (5.22) is again linear. This quadratic suppression of the deviation angle turned out to be sufficient to make the experiment of Ives and Stilwell work.

Ives and Stilwell used atomic hydrogen as moving emitters of electromagnetic radiation. More precisely, they used the second spectral line of the so-called 'Balmer series' (after the Swiss physicist and mathematician Johann Balmer (1825–1898)). This line is usually denoted by H_β and has a wavelength of 4861 Å, corresponding to a blueish-green colour. To produce hydrogen atoms in motion they first accelerated H_2^+ and H_3^+ ions by some voltage and then turned the ions into excited atomic hydrogen by neutralization and subsequent dissociation. Accordingly, the hydrogen atoms in the beam came in two different velocities of ratio $\sqrt{3}/\sqrt{2}$, depending on whether the hydrogen atom originated from an H_3^+ or H_2^+ ion. The accelerating voltage was varied in the interval between 6788 and 18 356 volts, producing β factors (velocities in units of c) of at most $4.4 \cdot 10^{-3}$ for the fast component stemming from H_2^+. At these velocities the transverse Doppler effect makes relative corrections of at most 10^{-5} or, in absolute terms, given the wavelength above, this corresponds to almost $5 \cdot 10^{-2}$ Å. In contrast, the longitudinal Doppler effect makes relative corrections of at most β, which in absolute terms corresponds to 21 Å. The factor by which the transverse Doppler effect is suppressed against the longitudinal one is $\beta/2 = 2.2 \cdot 10^{-3}$. Nevertheless, Ives and Stilwell achieved a relative accuracy of $5 \cdot 10^{-7}$, corresponding to $2.5 \cdot 10^{-3}$ Å in absolute terms.

The story acquires an ironic twist through the remark that Ives and Stilwell actually did not believe in SR. To be sure, they did believe that (5.18) was the right description of the Doppler effect, but they understood it in the context of the old ether theory of Lorentz and Larmor. This theory, together with the hypotheses that relative motion with respect to the ether causes length contractions and, in addition, time dilations, is operationally indistinguishable from SR. They do not differ in any predictions concerning observations. Rather they differ in their content concerning non-observable structures, i.e. the ether. Also, whereas length contraction and time dilation are clear consequences of unambiguous operational definitions of length and time in SR, they cannot be deduced in the Lorentz–Larmor theory, but have to be put in 'by hand'. This is simply because that 'theory' is merely a set of loosely connected assumptions without an actual basis. There is no real dynamical theory of some 'ether' from which conclusions concerning its interaction can be unambiguously deduced. Hence it has less explanatory power and, in that sense, cannot compete with SR.

5.6 The current experimental status of Special Relativity

So far we have encountered three historical experiments which relate SR to the world of physical phenomena. These were

- MM: The Michelson–Morley experiment, testing a possible dependence of the speed of light on the *direction* of the relative velocity with respect to a hypothetically preferred reference system K_0 (usually called the 'ether system' or 'ether frame').
- KT: The Kennedy–Thorndike experiment, testing a possible dependence of the speed of light on the *modulus* of the relative velocity with respect to a hypothetically preferred reference system K_0.
- IS: The Ives–Stilwell experiment, testing the time dilation of moving clocks.

In modern terminology one calls any experiment by one of these compound names, usually in the abbreviated form MM, KT, and IS, if it tests the associated aspects. This is done irrespective of whether the modern counterparts of these experiments bear any closer resemblance to the actual historic ones, which is often not the case.

A very important concept for the theoretical description of such experiments is given by *test theories*. Generally speaking, these are theories which allow one to incorporate and parametrize possible violations of the aspects and predictions in question. For this one needs to make some plausible assumptions about the nature of the expected deviations. Only those deviations will then be parametrized and tested. A test then results in constraints—usually upper bounds—on these parameters. More abstractly one may say that in the 'space of all possible theories' test theories parametrize certain (finitely many and hopefully cleverly chosen) directions which are then exposed to experimental scrutiny. Note that the experiments, then, make no statement about the other directions in 'theory space'. It is important to understand that *any* quantitative statement about the degree of validity of a theory is of that kind.

Applying this general idea to SR, it has been convincingly argued that any 'reasonable' violation of SR is parametrized by a three parameter family of test theories [27, 28]. Moreover, the parameters are uniquely fixed by the results of the three types of experiments mentioned above. It is for this reason that we focused attention on these three types. Special Relativity is precisely one member of this three-parameter family, corresponding to null results for the MM and KT experiments, and to the value (2.14) for the time dilation factor γ tested by the IS experiment. In this sense we can identify the experimental status of SR with the status of the MM, KT, and IS experiments. A fairly complete—though not quite up to date—list of the types of experiments that were performed in connection with SR can be found in [29].

The various test theories in our three parameter class differ, in particular, in their prediction of how the velocity of light depends on the observer's state of motion relative to the (hypothetically) preferred system. This aspect is labelled by two parameters, A and B. The general expression for the velocity of light is then written in the form:

$$c(v, \theta) = c_0 \left(1 + A \frac{v^2}{c_0^2} + B \frac{v^2}{c_0^2} \sin^2 \theta \right). \quad (5.23)$$

Here c_0 is the velocity of light in the preferred (ether-) system K_0. v and θ label the modulus and direction of the observer's velocity relative to K_0. Here a possible azimuthal dependence was excluded for simplicity: as long as no dependence on θ shows up in any experiment, no azimuthal dependence needs to be considered in the first place. A and B now parametrize this dependence on v and θ. $B = 0$ implies isotropy of the velocity of light. If, in addition, $A = 0$, then c is also independent of v. The MM and KT experiments therefore set upper bounds to B and A respectively. Quite generally, KT experiments turn out to be less accurate than MM experiments. In modern versions the estimated relative errors in KT experiments are larger by at least two orders of magnitude. This is basically a direct consequence of the fact that during a KT experiment the relative change in the modulus of the velocity (5.10) is much less than is the relative change of its direction, measured, e.g. by $\cos\theta$, during a MM experiment.

Technological progress in material sciences and laser physics have recently stimulated new and powerful MM, KT, and IS experiments of greatly enhanced precision. Currently the best MM and KT experiments give (see [30] for MM and [31] for KT; the results are to be understood at the level of one standard deviation, i.e. 84% confidence level)

$$\text{MM:} \quad \frac{\Delta c}{c_0} = 4.3 \cdot 10^{-15} \Rightarrow |B| < 3.7 \cdot 10^{-9}, \quad (5.24)$$

$$\text{KT:} \quad \frac{\Delta c}{c_0} < 1.6 \cdot 10^{-12} \Rightarrow |A| < 10^{-6}. \quad (5.25)$$

Here Δc is the variation of c which derives from the measured variation of frequencies. Using (5.23) then allows one to deduce the given upper bounds for the moduli of A and B, under the assumption that $v = 380$ km/s. We cannot give a fair account of these high-technology experiments at this point. Let it merely be mentioned that the interferometer of MM and KT is now realized on a much smaller scale, using cryogenic optical resonators, which are throughout kept at the temperature of liquid helium (-269 degrees Celsius). Refined laser techniques produce highly stable resonance frequencies and hence a long-term stability of the physical conditions. In fact, the stability is so good that the experimenters decided not to let any device rotate the interferometer in their laboratory, which will always cause mechanical disturbances, but rather to wait a couple of hours until its orientation (with respect to the hypothetical preferred frame) had been changed by the Earth's own rotation. See [32] and [33] for more information on their experiments.

The third parameter of our test theories parametrizes the possible deviations of the factor γ of time-dilation from that given by (2.14). Hence one replaces γ according to

$$\gamma \to \gamma(1 + \alpha(\beta^2 + 2\vec{\beta}_0 \cdot \vec{\beta})). \tag{5.26}$$

Here $\vec{\beta} = \vec{v}/c$ and $\vec{v}$ is the velocity of the moving clock relative to the observer and $\vec{\beta}_0 = \vec{v}_0/c$, where $\vec{v}_0$ is the velocity of the observer relative to the ether system. Currently the best result is given by [34]

$$|\alpha| < 2.2 \cdot 10^{-7}. \tag{5.27}$$

Let us say a few words on this fascinating modern version of the Ives–Stilwell experiment, performed at the heavy-ion storage ring of the Max-Planck-Institute at Heidelberg (Germany). The moving 'clocks' are atomic transitions in singly-ionized atoms of lithium, $^7Li^+$, accelerated to an average speed of 19 000 kilometres per second, which corresponds to 6.3% of the velocity of light. The method employed may be called 'high resolution saturation spectroscopy', for reasons explained below.

Simple Doppler spectroscopy would consist of tuning a laser to be in resonance with a certain two-level atomic transition of the moving ion. Let the resonance frequency, as measured in the ion's rest frame, K', be ν_0 (here we prefer to write ν_0 rather than ν', which would be suggested by our systematics.). At resonance the laser induces a transition to the excited state. Subsequently the ion returns to the ground state under emission of light, a process which we here wish to refer to as fluorescence (by some abuse of terminology). It is the occurrence of this characteristic fluorescence light, which is verified by photomultipliers, that signals that the resonance frequency has been found. Now let the ion (system K') move relative to the laboratory (system K) at velocity v in the x direction. Let the laser beam also point in the same direction, i.e. following the ion. Then the laser frequency, as measured in the laboratory, has to be tuned above the resonance frequency, since the light, as seen by the ion, suffers a Doppler shift toward lower frequencies. More precisely, according to (3.10) (setting $v' = v_0$), the laser has to be tuned to $v = v_0 \gamma (1 + \beta)$. This is schematically shown in **Fig. 5.5**. Knowing the resonance frequency and the velocity of the ion, γ can be experimentally measured.

But this is *not* what is done in the experiment of the Heidelberg group, the reason being that particle velocities are much harder to control than laser frequencies. Hence the idea is to measure a velocity-independent quantity. In saturation spectroscopy, this is achieved as follows: We again consider the two-level system, but now two lasers are used to excite the ions in the beam. One laser points parallel and the other anti-parallel to the direction of the ion beam, as schematically pictured in **Fig. 5.6**. The line

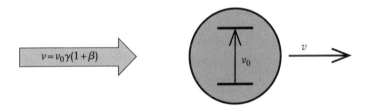

Fig. 5.5 Simple Doppler spectroscopy.

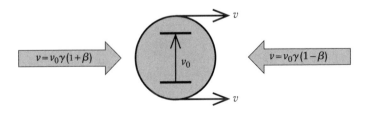

Fig. 5.6 Double Doppler spectroscopy.

width of the lasers is only 1/3000 of the Doppler width of the velocity distribution of the ions (1 MHz versus 3 GHz). Hence the two lasers will generally be in resonance with ions from different, non-overlapping velocity classes. The mean beam-velocity is now so adjusted to have a maximal fluorescence yield from, say, the parallel laser, which is of fixed frequency. This makes sure that the parallel laser is in resonance with the ions around the centre of the velocity distribution. In addition, the laser's intensity is turned up to reach saturation in ion excitations. This means that any further increase of its intensity will not significantly increase the fluorescence yield. Now the other laser, which is tunable in its frequency, is switched on. Initially, its frequency, as seen by the ions, does not match that of the first laser. Hence it will excite ions from a different velocity class within the beam. The fluorescence yields from both lasers now simply add, since the light comes from different ions. The second laser's intensity is also held at saturation, so that their common fluorescence yield is about twice that of any one of them. Now the second laser's frequency is tuned to excite the *same* velocity class of ions as the first laser. This frequency is found by looking for a dip in the fluorescence yield. The dip occurs because both lasers are at saturation intensity and now share the *same* ions. Hence the fluorescence yield drops back to a level not significantly higher than that of a single laser. This is depicted in **Fig. 5.7**. Let the laser frequencies, as measured in the laboratory (System K), be ν_1, for the parallel pointing laser, and ν_2 for the anti-parallel pointing one. That both are in resonance with the atomic transition of frequency ν' in the rest system of the ion means that $\nu_1 = \nu_0 \gamma (1 + \beta)$ and $\nu_2 = \nu_0 \gamma (1 - \beta)$, according to (3.10) (again writing ν_0 for ν'). Hence

$$\frac{\nu_1 \cdot \nu_2}{\nu_0^2} = \gamma^2 (1 - \beta^2) \approx (1 + 2\alpha(\beta^2 + 2\vec{\beta}_0 \cdot \vec{\beta})), \tag{5.28}$$

where in the last step we replaced γ by the right-hand side of (5.26), neglecting terms of higher than linear order in α. The numerator of the quotient on the left side of this equation is the desired velocity-independent quantity that is actually

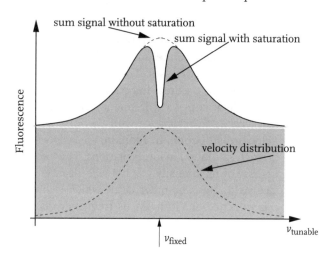

Fig. 5.7 Dip in fluorescence yield at double resonance in saturation spectroscopy.

measured. The rest-system transition frequency, ν_0, is known from independent measurements up to relative errors of about $2 \cdot 10^{-9}$, which gives the denominator. Note also that the quadratic term in β dominates the right-hand side of (5.28), as long as the ion velocity, here given by $\beta \cdot c = 6 \cdot 3 \cdot 10^{-2} \cdot c$, is considerably larger than the (assumed) velocity $\beta_0 \cdot c$ against the ether system. This is certainly the case if the latter is identified with the rest system of the cosmic background radiation, where $\beta_0 \cdot c = 380 \, \text{km/s} = 1.3 \cdot 10^{-3} \cdot c$. Making this assumption, measurements [34] of the left hand side of (5.28) led to (5.27). More information on the experimental setup may be found at [35].

Let us also point out that there is also a certain fundamental aspect in the coincidence of the factor of time dilation with $\gamma = 1/\sqrt{1-\beta^2}$. It can be shown that this is precisely the necessary and sufficient condition for Einstein synchrony to be identical with that given by slow clock-transport; see [28]. In particular, these two synchronization prescriptions are compatible within SR. This will be explained in the next section.

5.7 Synchronization by slow clock-transport

In Sect. 3·1 we already mentioned the possibility of synchronizing the clocks in some inertial system by means of a master clock, U_T. That clock would be moved through all of space such that, at each point, the local clock can be synchronized with it. Since this is obviously a rather impractical way of doing things, we decided for Einstein's synchronization based on the exchange of light signals. On the other hand, clock transport has the advantage of being conceptually rather simple and clear cut, since it does not rely on much more than the existence of clocks (and a moving agent), whereas Einstein's definition brings in the laws of light propagation in the limiting case of geometric optics. In this section we will show that, as far as SR is concerned, these definitions are equivalent.

Let us therefore consider a clock U_T initially at rest with respect to the inertial system K'. We assume the clocks of K' to be Einstein-synchronized, thus giving rise to the globally defined time t' of K'. Now we start to move U_T along the x'-axis with velocity u' relative to K'. We adjust U_T so that it shows time zero at $x' = 0$ and $t' = 0$. Due to time dilation, U_T's reading is retarded compared to time t' by a factor of $1/\gamma(u')$. (Recall that γ is always understood as a function of some velocity, as introduced in (2.14). Since in this section we need to consider the γ-factors for various velocities at the same time, we shall explicitly write down the arguments to prevent confusion.) In order to get from $x' = 0$ to $x' = l'$, the clock U_T needs the t'-time l'/u'. Hence its own reading at $x' = l'$ is $l'/(u'\gamma(u'))$, which makes a retardation of

$$\frac{l'}{u'} \cdot \left(1 - 1/\gamma(u')\right) \approx \frac{u'l'}{2c^2}. \tag{5.29}$$

Here the $\approx$ stands for equality up to powers in u' higher than the first (linear approximation). We see that the discrepancy between the local t'-time and the reading of U_T can be made arbitrarily small by choosing the transport velocity u' of U_T. This is sometimes expressed by saying that U_T reads the time t' of K' in the limit of 'infinitely small' transport velocities relative to K'. This is what is meant by 'slow transport' in the header of this section.

What needs to be shown is that the last statement is shared by *every* inertial observer, i.e. that it satisfies the relativity principle. Let us investigate concretely what this entails. Suppose K is another inertial system relative to which K' moves with velocity v in the x direction. The clocks of K are also assumed to be Einstein-synchronized, thus giving rise to the global time t. As usual, we choose the axes of K and K' to coincide for $t = t' = 0$. The velocity u of U_T relative to K is given by the addition law (3.6):

$$u = \frac{u' + v}{1 + u'v/c^2} \approx v + u'\gamma^{-2}(v). \tag{5.30}$$

Again $\approx$ denotes the linear approximation in u'. At time t the clock U_T is at position $x = ut$ relative to K. At this moment its reading is given by

$$t/\gamma(u) \quad \text{reading of } U_T, \tag{5.31}$$

as follows from time dilation. On the other hand, using the Lorentz transformations (3.3), we can calculate the time t' of K' that corresponds to time t and position ut relative to K:

$$t' = \gamma(v)(t - vut/c^2) = t\,\gamma(v)(1 - vu/c^2). \tag{5.32}$$

The non-trivial requirement now is that this equals the clock's reading (5.31) in the linear approximation in u'. Equating these expressions then shows that we must have

$$\gamma^{-1}(u) \approx \gamma(v)(1 - vu/c^2), \tag{5.33}$$

where u on both sides stands for the expression in (5.30). Expanding both sides of (5.33) to linear order in u' indeed shows equality.

Hence all inertial observers agree that the two synchronization procedures in K' lead to the same notion of time t' in K'. This is by no means obvious and tied to specific properties of the Lorentz transformation, which entered explicitly on the right-hand side of (5.33), and also on the left-hand side through its expression for time dilation. In fact, as already mentioned, it has been shown in [28] that a necessary and sufficient condition for the equivalence of slow

5.8 Aberration and conformal transformations

In Sect. 3.6 we discussed the special relativistic law of aberration. There we found that the functional relation between the angles assumes its simplest form if expressed in terms of the tangents of half the angles, as stated in (3.9). This fact has a deeper mathematical interpretation which we now wish to explain. This interpretation will then immediately lead to a general proof of the fact that the visual image of a fast moving spherical body is again spherical and not contracted, as was already stated in Sect. 3.7.

Figure 5.8 is the schematic representation of an observer located at *B*, who receives a light ray (dashed line) at an angle of

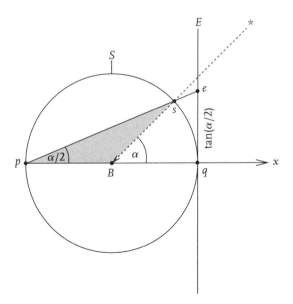

Fig. 5.8 Aberration as conformal transformation.

incidence, α, with respect to the x-axis. The observer is at the centre of an imaginary sphere, S, sometimes called the 'observer sky' of B, the diameter of which is chosen to be of unit length. Every light ray received by the observer intersects this sphere at some unique point. Moreover, to every point of S we can uniquely associate a light ray whose point of intersection is the given one. Hence the association of light rays received by B with points on S is a unique and invertible one. In technical terms, it is a bijection. For example, in **Fig. 5.8**, the given light ray (dashed line) corresponds to the point s. Abstractly speaking, we may identify the set of all light rays received by B with the set of points on S. This is quite analogous to what happens in a planetarium, where our natural visual impression of stars and planets is imitated by light spots projected on the hemispherical ceiling.

We will now introduce an alternative projective representation, the so-called 'stereographic projection', which will lead to the sought for interpretation. We start by selecting some antipodal pair of points p and q on S; see **Fig. 5.8**. At q we attach the plane E that is tangential to S. We can now project any point s of S, other than p, on to E. This we do by letting the image point of the projection, s', be the unique intersection point of the line through p and s with E. This prescription defines a unique map from all points of S, except p, to E. Conversely, it is obvious that any point of E is the image of some uniquely determined point on S (without p). This map from S to E is called the stereographic projection of S with projection centre p. The stereographic projection allows one to uniquely identify all but one light rays received by B with points on E. The one exceptional light ray, which is not represented on E, is that which meets S at p, i.e. where the angle of incidence α is 180°.

Let us now choose q, the antipodal point of p, as the origin of E. The crucial observation is now that a light ray of incidence angle α corresponds, via stereographic projection, to a point on E whose distance to the origin is just $\tan(\alpha/2)$. This follows immediately from **Fig. 5.8** and elementary Euclidean geometry. Indeed, since the segments $\overline{pB}$ and $\overline{sB}$ are equally long, the angles of the shaded triangle at p and s must also be equal. Hence their sum equals twice the angle at p, which in turn must be equal to 180° minus the angle at B. The latter difference is obviously α, hence the angle at p is $\alpha/2$, as already anticipated in **Fig. 5.8**. Recalling that the diameter of S had been chosen to be the unit length, this shows that $\overline{eq}$ has length $\tan(\alpha/2)$, as was to be shown.

We now consider an observer B' moving relative to B at velocity v along the x-axis. At time $t = 0$ both observers are at the origin of S. At this particular moment we can do all constructions mentioned so far in connection with observer B also for the second observer B'. The angle of incidence measured by B' will be denoted by α'. The quantities $\tan(\alpha/2)$ and $\tan(\alpha'/2)$ are now the distances in the common plane of both stereographic projections. The law of aberration (3.9) now simply states that these distances are scaled by the factor $\sqrt{(1-\beta)/(1+\beta)}$. This scaling map is a linear radial contraction (for $\beta > 0$; for $\beta < 0$ it is a dilation). This is what the law of aberration boils down to in the stereographic-projection representation.

Scaling maps are similarity transformations. In particular, they preserve angles. Generally a transformation is called 'conformal' if it preserves angles, even if it is not linear. For example, the stereographic projection is conformal, meaning that any two curves on S, which intersect at some angle, are mapped to curves on E intersecting at the same angle. Moreover, the stereographic projection has the following nice property, whose proof we omit: it maps circles on S not intersecting p to circles on E; circles intersecting p are mapped to straight lines. The converse of this is also true. It follows that discs on S not containing p are mapped to discs on E. This is because the boundary of a disc κ_S on S is a circle, whose image on E is then also a circle. But the latter circle must be the boundary of κ_E, the image of κ_S under the stereographic projection. Hence κ_E is also a disc.

Using this we can immediately deduce that spherically shaped bodies in motion do not appear contracted but still spherical, as was already announced in Sect. 3.7. The argument is this: consider a spherically shaped body at rest with respect to observer B. Light rays originating at the body and ending at B intersect S in a round disk, κ_S, of circular boundary. That disc represents the two-dimensional image actually seen by the observer. We wish to prove that B' also sees a round rather than a squashed disc. For this we simply apply the law of aberration in the form given above. So let κ_E be the image of κ_S on E under stereographic projection. As we just explained, it is also a disc. Now apply to κ_E the scaling map corresponding to aberration. Let the result be κ'_E. This, too, will be a disc, since scalings map circles to circles and hence discs to discs. Finally apply the inverted stereographic projection to κ'_E. The result is a disc κ'_S on S. By construction, the map $\kappa_S \to \kappa'_S$ corresponds to the law of aberration. Since it is a composition of conformal maps is it itself conformal. It maps the visual impression of B to that of B'. This shows that B', too, sees a disc.

5.9 Transformation formulae for momentum, energy, and force

In Sect. 3.2 we explicitly stated how space and time coordinates transform under the special Lorentz transformation corresponding to a boost in the x direction. The relevant formulae were presented in (3.3). We also deduced the corresponding transformation rules for velocities, given in (3.6). In this section we wish to complete these by deriving the transformation laws for momentum, energy, and force.

We recall the expression (3.16) for momentum and (3.24) together with (3.15) for the total energy of a moving body of rest mass m_0. We assume the body moves at velocity $\vec{u}'$ relative to an inertial system K'. Its momentum and energy relative to K' are then given by

$$\vec{p}' = m_0 \gamma(u') \vec{u}' \quad \text{and} \quad E' = \gamma(u') m_0 c^2 \qquad (5.34)$$

respectively. (As usual, u' denotes the modulus of $\vec{u}'$.) Let K be an inertial system, relative to which K' moves at velocity v along the x-axis. Using the formulae (3.6) for the addition of velocities, we can immediately determine the velocity $\vec{u}$ of the body relative to K. Direct computation shows that the γ-factor for this combined velocity satisfies

$$\gamma(u) = \gamma(u')\gamma(v)(1 + u'_x v/c^2). \tag{5.35}$$

Relative to K, the momentum and energy of the body are given by

$$\vec{p} = m_0 \gamma(u) \vec{u} \quad \text{and} \quad E = \gamma(u) m_0 c^2 \tag{5.36}$$

respectively. Using (5.35), they can be re-expressed in terms of $\vec{p}'$ and E' as given by (5.34). This leads directly to the sought for transformation rules (for brevity, we now write again $\gamma(v) = \gamma$):

$$p_x = \gamma(p'_x + vE'/c^2), \qquad p_y = p'_y, \qquad p_z = p'_z,$$
$$E = \gamma(E' + vp'_x). \tag{5.37}$$

Comparing (5.37) with the inverse of (3.3) shows that the tuples $(E/c, \vec{p})$ and $(ct, \vec{x})$ transform in the same fashion under boost transformations. This is clearly also true for spatial rotations, but not for space-time translations, under which E and $\vec{p}$ stay invariant. Moreover, using (5.37), one easily shows that

$$E^2/c^2 - \vec{p} \cdot \vec{p} = E'^2/c^2 - \vec{p}' \cdot \vec{p}'. \tag{5.38}$$

Hence the value of these expressions is invariant under the special Lorentz transformation considered here (boost in the x direction). But it is easily seen to be also invariant under spatial rotations (under which the energy and the squared momentum are separately invariant) and under space-time translations (under which energy and momentum—without squaring—are also separately invariant). Hence the difference between E^2/c^2 and the squared momentum is a fully Lorentz invariant quantity. To determine its value we might just evaluate it in the system where the momentum vanishes (centre-of-mass system). There its energy is just $m_0 c^2$ (cf. (3.24)), so that the value of the expressions in (5.38) is given by $m_0^2 c^2$. This is essentially what we already found in (3.25).

Let us now turn to the transformation law for the force. Here we recall that the Newtonian force relative to K and K' is given by the rate of change of momentum $\vec{p}$ and $\vec{p}'$ with respect to time t and t', respectively. The increment dt of time t can be expressed through the increment dt' of time t' by using the Lorentz transformation, $t = \gamma(t' + vx'/c^2)$. Here x' is the location of the body in K', so that we have to set $x' = u'_x t'$. This leads to $dt = \gamma(1 + vu'_x/c^2)dt'$. Finally we recall that, by the principle of energy conservation applied in system K', the change dE' of the energy E' in the time interval dt' must equal the work done by the force $\vec{F}'$ during dt'. This gives $dE'/dt' = \vec{F}' \cdot \vec{u}'$. Now, taking the time derivative of the formulae (5.37) for momentum immediately implies the following transformation rules for the (Newtonian) forces

$$F_x = \frac{F'_x + \frac{v}{c^2} \vec{F}' \cdot \vec{u}'}{1 + vu'_x/c^2}, \quad F_y = \frac{F'_y}{\gamma \left(1 + vu'_x/c^2\right)}, \quad F_z = \frac{F'_z}{\gamma \left(1 + vu'_x/c^2\right)}. \tag{5.39}$$

As usual, the inverse relations of (5.37) and (5.39) are obtained by exchanging primed and unprimed quantities and at the same time replacing v by $-v$.

5.10 Minkowski space and the Lorentz group

In Sect. 3.3 we have learned certain rules about how to determine separation lengths of pairs of events in space and time. For example, given two events O and E, such that E is in the causal complement of O, like e.g. in **Fig. 3.10**, we can find an inertial reference system, K', in which O and E are simultaneous (happen at the same time t' of K'). Hence we can define the *space-time distance* between O and E by their spatial distance in K'. Likewise, if E is in the chronological past or future of O, i.e. can be reached from O by subluminal propagation, then we can find a system K' with respect to which O and E are equilocal (happen at the same spatial position (x', y', z') of K'). In this case we can define the *space-time distance* between O and E through the time separation (times c) in K'.

Minkowski space and the Lorentz group

The rules for calculating these distances were stated in Sect. 3.3. However, a simple and far reaching observation allows us to do this in a far more elegant way. This we shall explain below. It is due to the mathematician Hermann Minkowski [11], whom we already encountered in connection with causality structures in Sect. 3.5. Minkowski's idea was to endow *space-time* with a generalization of what mathematicians call a *distance function*, or a *metric* for short. Let us briefly recall a few notions concerning metrics.

Quite generally, a metric is a real-valued function $d(p, q)$ of two arguments, where p and q denote two points of the space that is to be endowed with this metric d. The number $d(p, q)$ is to be interpreted as the distance between p and q. Now, traditionally, a metric has to satisfy the following three axioms:

1. $d(p, q) = d(q, p)$, for all p and q. This is called the symmetry of d.
2. $d(p, q) = 0$ if and only if $p = q$.
3. $d(p, r) + d(r, q) \geq d(p, q)$ for all p, r, and q. One says that d obeys the triangle inequality. (Think of p, r, and q as the vertices of a triangle and of $d(p, q)$, for example, as the length of the edge joining p and q.)

For example, the three-dimensional real vector space $\mathbb{R}^3$ is usually endowed with the familiar 'Euclidean metric', given by

$$d(\vec{x}_1, \vec{x}_2) = \sqrt{(\vec{x}_1 - \vec{x}_2) \cdot (\vec{x}_1 - \vec{x}_2)}$$
$$= \sqrt{(x_1 - x_2)^2 + (y_1 - y_2)^2 + (z_1 - z_2)^2}, \qquad (5.40)$$

where $\vec{x}_1 = (x_1, y_1, z_1)$ and $\vec{x}_2 = (x_2, y_2, z_2)$. This metric clearly satisfies the three axioms above. Note that the Euclidean metric is invariant under spatial translations:

$$\vec{x} \mapsto \vec{x} + \vec{a}, \qquad (5.41)$$

and spatial rotations

$$\vec{x} \mapsto \mathbf{R} \cdot \vec{x}. \qquad (5.42)$$

Here $\vec{a}$ is a fixed translation vector and $\mathbf{R}$ is a so-called 'orthogonal' 3×3 matrix. In algebraic terms, orthogonality means that the transposed matrix $\mathbf{R}^\top$ is the inverse of $\mathbf{R}$, i.e. $\mathbf{R}^\top \cdot \mathbf{R} = \mathbf{1}$, where $\mathbf{1}$ is the unit matrix. This is identical to the condition that orthogonal transformations leave the Euclidean scalar product invariant. This means that the $\mathbf{R}$-transformed vectors have the same scalar product as the untransformed ones. For later comparison, we note this condition in the following form:

$$\mathbf{R}^\top \cdot \mathbf{1} \cdot \mathbf{R} = \mathbf{1}. \tag{5.43}$$

It is not hard to show that (5.41) together with (5.42) are the only inhomogeneous linear (inhomogeneous meaning that translations are included) transformations of $\mathbb{R}^3$ that preserve the Euclidean distance (5.40). (In fact, a much stronger result holds, namely that *any* map of $\mathbb{R}^3$ to itself that preserves Euclidean distance, is of the form stated. Note that here neither affine linearity nor even continuity need to be assumed a priori. This follows from a famous theorem due to Beckman and Quarles [36].) This means that the Euclidean distance function fully characterizes all those transformations which can be obtained by composing translations with rotations. They are therefore called 'Euclidean motions'.

Now we return to space-time. As stated above, Minkowski's idea was to endow four-dimensional space-time (not just three-dimensional space) with some distance function, that could be used to fully characterize Lorentz transformations in a way similar to the characterization of Euclidean motions by the Euclidean distance function. In fact, we have already given the prescription for this distance function above. It assigns the simultaneous spatial distance in case the points are in the causal complement of each other, and the equilocal time separations (times c) in case the points are in the chronological future and past of each other. What remains to do is to bring this definition into a more convenient form. In fact, as Minkowski observed, its mathematical expression in any inertial reference system turns out to be surprisingly simple. To see this, let $(ct_1, \vec{x}_1)$ and $(ct_2, \vec{x}_2)$ be the coordinates of O and E with respect to the system K, and likewise $(ct'_1, \vec{x}'_1)$ and $(ct'_2, \vec{x}'_2)$ the coordinates of the same events with respect to K'. Let, as usual, K'

Minkowski space and the Lorentz group

be moving with respect to K at speed v in the x direction. Then the primed coordinates can be expressed in terms of the unprimed ones by the Lorentz transformation as given in (3.3). From this it is easy to prove that

$$c^2(t_1 - t_2)^2 - (\vec{x}_1 - \vec{x}_2) \cdot (\vec{x}_1 - \vec{x}_2)$$
$$= c^2(t'_1 - t'_2)^2 - (\vec{x}'_1 - \vec{x}'_2) \cdot (\vec{x}'_1 - \vec{x}'_2). \qquad (5.44)$$

This difference of squares is obviously also invariant under space-time translations, where $t \mapsto t + b$ for some real number b and $\vec{x}$ transforms as in (5.41). Moreover, it is invariant under spatial rotations as in (5.42), which leave time invariant. But any general inhomogeneous Lorentz transformation is a composition of a boost in the x direction, a spatial rotation, and a space-time translation. Hence the expression (5.44) is, in fact, invariant under *all* general inhomogeneous Lorentz transformations.

Now, if O and E are in the causal complements of each other, and the system K' is such that $t'_1 = t'_2$, the right-hand side of (5.44) is just minus the square of the spatial distance in K'. If, on the other hand, O and E are in the chronological future and past of each other, we can choose K' such that $\vec{x}'_1 = \vec{x}'_2$ and the right-hand side of (5.44) is just the square of the time distance of the two events in K'. Together this implies that the space-time distance measure defined at the beginning of this section can be expressed in any system K in the following simple form:

$$d(O, E) = \sqrt{|c^2(t_1 - t_2)^2 - (\vec{x}_1 - \vec{x}_2) \cdot (\vec{x}_1 - \vec{x}_2)|}. \qquad (5.45)$$

Expression (5.45) is usually referred to as the 'Minkowskian distance' between O and E. It looks almost like a straightforward generalization of the Euclidean distance function (5.40) to four dimensions, were it not for the crucial relative minus sign (instead of plus) between the squares of the time and space intervals. For this reason we had to take the modulus of the quantity under the square root, since otherwise it may become negative. Moreover, whereas this new distance function is also symmetric, $d(O, E) = d(E, O)$, it ceases to satisfy the usual axioms 2. and 3. stated above. So the generalizations allowed by Minkowski indeed do make a big difference. For example, for any event E on the light cone of O the Minkowskian distance according to (5.45) is zero. Furthermore, consider a triangle in space-time two sides of which are segments of light rays. Then these sides have zero Minkowskian length, in clear violation of the triangle inequality.

Since taking the modulus of a quantity erases the information about its sign, it is more appropriate to consider the square of the distance function (5.45) without taking its modulus. Hence we define

$$D(O, E) = c^2(t_1 - t_2)^2 - (\vec{x}_1 - \vec{x}_2) \cdot (\vec{x}_1 - \vec{x}_2). \tag{5.46}$$

This is usually called the Minkowskian squared distance, though this is slightly misleading, since, having dropped the moduli signs, it is not the square of any real number (it may assume negative values). Nevertheless, D is a useful object to consider, since its values now reveal certain information on the causal relations of its arguments. Except for the trivial case where $O = E$, there are three further cases to be distinguished:

1. $D(O, E) < 0$; it follows that E is in the causal complement of O and O in the causal complement of E. One says that the points O and E are space-like separated.
2. $D(O, E) > 0$; it follows that E is either in the chronological future or in the chronological past of O. Likewise, O is in the chronological past or future of E, respectively. One says that points O and E are time-like separated.
3. $D(O, E) = 0$ and $O \neq E$; it follows that E is either on the future or past light cone of O. Likewise, O then lies on the past or future light cone of E, respectively. One says that O and E are light-like separated.

Let O be a given point which we choose as origin of the coordinate system. We consider all points E for which $D(O, E) = 1$. Its coordinates are $(ct, \vec{x})$, where $\vec{x}$ stands for (x, y, z). We write r^2 for $\vec{x} \cdot \vec{x}$. Then the coordinates satisfy

$$ct = \pm\sqrt{r^2 + 1}. \tag{5.47}$$

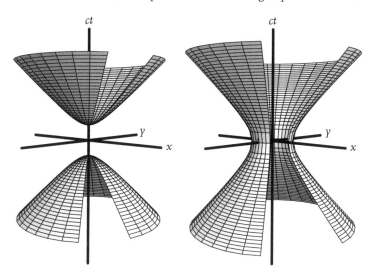

Fig. 5.9 Hyperboloids of all points of unit time-like Minkowskian distance (left) and unit space-like distance (right) to the origin. The slits are just for better visual impression.

This defines a two-sheeted hyperboloid, both sheets of which comprise all points of unit time-like Minkowskian distance to the origin. The left picture of **Fig. 5.9** represents them in a three-dimensional space-time version, where we suppressed one space dimension (z-coordinate). The upper sheet corresponds to the plus sign in (5.47), consisting of all points which lie in the chronological future of the origin, and the lower sheet corresponds to the minus sign, where all points lie in the chronological past of the origin. The first case corresponds to the upper curve of **Fig. 3.4**, after suppression of one more spatial dimension (y-coordinate).

Next we consider all events E for which $D(O, E) = -1$. E's coordinates now satisfy

$$r = \sqrt{c^2 t^2 + 1}. \qquad (5.48)$$

It describes a one-sheeted hyperboloid, all points of which are space-like separated a unit length away from the origin O. It is shown in the right picture of Fig. **5.9**, where again one spatial dimension (z-axis) is suppressed.

Together, these hyperboloids of **Fig. 5.9** are the analogues in Minkowskian geometry of the single unit-sphere in Euclidean geometry. We have seen above that Euclidean motions preserve the Euclidean distance and that Lorentz transformations preserve the Minkowski distance. We also mentioned that the Euclidean motions are the *only* transformations of space that preserve the Euclidean distance. Now, this is also true for the Lorentz transformations. In fact, any linear inhomogeneous transformation that preserves the Minkowski distance (or, equivalently, its square (5.46)) must be a composition of the following transformations: a boost in the x direction (3.3), a spatial rotation (5.42), and a space-time translation. Note that in this case it is well motivated to a priori assume the transformations to be linear inhomogeneous, since these are the most general ones that transform straight lines, representing inertial motion, to straight lines. But for completeness we also mention that, as in the Euclidean case, we may significantly strengthen the mathematical statement. In fact, any bijection of $\mathbb{R}^4$ (space-time) that, together with its inverse, preserves the Minkowskian distance (here it would even suffice to restrict either to time-like or space-like distances) must be a composition of the transformations just listed. This is a consequence of a famous theorem of Alexandrov's; see e.g. [37] for a comprehensive account. It seems to be the closest analog in Minkowskian geometry to the theorem of Beckman and Quarles in Euclidean geometry, though the latter did not require the hypothesis of bijectivity.

A purely mathematical treatment of SR most conveniently starts with the Minkowskian geometry and motivates the Lorentz transformations in the fashion just described. A powerful vector calculus can then be built on this structure, that resembles, to a certain extent, the vector calculus in $\mathbb{R}^3$. With respect to an inertial frame, a point in space-time can then be associated with a four component vector $X = (ct, \vec{x})$, simply called a 'four-vector'. A scalar product, traditionally denoted by the Greek letter η, between pairs

Minkowski space and the Lorentz group

of four-vectors can then be defined as follows:

$$\eta(X_1, X_2) = c^2 t_1 t_2 - \vec{x}_1 \cdot \vec{x}_2. \tag{5.49}$$

It is readily shown to be invariant under Lorentz boosts (3.3) and spatial rotations, and hence under all linear Lorentz transformations. If $\mathbf{\Delta}$ denotes the 4×4 diagonal matrix $\mathrm{diag}(1, -1, -1, -1)$, this invariance under a Lorentz transformation, given by the 4×4 matrix $\mathbf{L}$, takes the form

$$\mathbf{L}^\top \cdot \mathbf{\Delta} \cdot \mathbf{L} = \mathbf{\Delta}. \tag{5.50}$$

This is the analogue in Minkowskian geometry of the orthogonality relation (5.43) in Euclidean geometry. Note that the only difference is that the unit matrix $\mathbf{1}$ gets replaced by the matrix $\mathbf{\Delta}$. One may then say that the general inhomogeneous Lorentz transformation is given by

$$X \mapsto \mathbf{L} \cdot X + A \tag{5.51}$$

where $\mathbf{L}$ obeys (5.50) and where $A = (b, \vec{a})$ is some four-vector for the space-time translation. The square of the Minkowski distance between events O and E, which are represented by the four vectors X_1 and X_2 with respect to some inertial reference system, is then given by

$$D(O, E) = \eta(X_1 - X_2, X_1 - X_2). \tag{5.52}$$

This equation makes sense, since the right-hand side is invariant under all simultaneous transformations of X_1 and X_2 according to (5.51). Therefore it only depends on the space-time points O and E and not on the way they are coordinatized.

We shall write $(A, \mathbf{L})$ for a transformations of the form (5.51). Such transformations can clearly be composed. We write composition as simple juxtaposition with the transformation on the right as the one that acts first. For example, a simple calculation shows that composing $(A_2, \mathbf{L}_2)$ (acting first) with $(A_1, \mathbf{L}_1)$ (acting second) gives

$$(A, \mathbf{L}) = (A_1, \mathbf{L}_1)(A_2, \mathbf{L}_2) = (A_1 + \mathbf{L}_1 \cdot A_2, \mathbf{L}_1 \cdot \mathbf{L}_2). \tag{5.53}$$

Now recall that mathematicians call any set G a 'group', if there exists an operation, called 'multiplication', that assigns to any ordered pair (g_1, g_2) of elements from G their product, $g_1 g_2$, such that the following axioms are satisfied:

G1: The multiplication is associative, i.e. $g_1(g_2 g_3) = (g_1 g_2)g_3$ for all g_1, g_2, g_3 in G.

G2: There exists a (necessarily unique) so-called neutral element e in G, such that $eg = ge = g$ for all g in G.

G3: For each element g there exists a (necessarily unique) so-called inverse element g^{-1} in G such that $gg^{-1} = g^{-1}g = e$.

It is straightforward to check that the set of Lorentz transformations (5.51) form a group if multiplication is defined to be composition, as in (5.53). To see this, first note that the matrix $\mathbf{L} = \mathbf{L}_1 \cdot \mathbf{L}_2$ satisfies (5.50) if $\mathbf{L}_1$ and $\mathbf{L}_2$ separately do: write down (5.50) for $\mathbf{L}_1$ and multiply this relation with $\mathbf{L}_2$ from the right and $\mathbf{L}_2^\top$ from the left. Hence composition is an operation in the set of inhomogeneous Lorentz transformations. Furthermore, associativity follows from the associativity of matrix multiplication, the neutral element is $(0, \mathbf{1})$, and the inverse of $(A, \mathbf{L})$ is $(A, \mathbf{L})^{-1} = (-\mathbf{L}^{-1} \cdot A, \mathbf{L}^{-1})$. Note that $\mathbf{L}^{-1}$ also satisfies (5.50) if $\mathbf{L}$ does. To see this, take the inverse and then the transpose of (5.50), and use the fact that $\mathbf{\Delta}$ is invariant under this operation. This proves that the set of all transformations (5.51), where $\mathbf{L}$ obeys (5.50), form a group under composition. This group is called the 'general inhomogeneous Lorentz group'. It has two obvious subgroups. One is the group of all translations, given by the set of elements of the form $(A, \mathbf{1})$. The other is the group of general homogeneous Lorentz transformations, given by the set of all elements of the form $(0, \mathbf{L})$ where $\mathbf{L}$ satisfies (5.50). These two groups work together in the multiplication law (5.53), which defines what is called the 'semi-direct product' of the translation group with the general homogeneous Lorentz group. The group property of Lorentz transformations was first established by Poincaré.

Minkowski's geometric formulation lies at the heart of modern technical presentations of SR. It elegance and power are unsurpassed. But is should not be forgotten that this 'geometry of space-time' is to be understood as 'physical geometry'. This means that in the first place it refers to spatial-temporal relations of real physical systems (objects), and not to abstract mathematical points. It is not a priori given to us, but derives from the dynamical rules (equations of motion) that these physical systems obey. This was how the Lorentz group was found in the first place: as a symmetry group of Maxwell's equations. Here one may recall the opening words of Minkowski's famous address to the 80th Assembly of German Natural Scientists and Physicians, where he, for the first time, presented his ideas to a wider audience [11]:

Gentlemen! The views of space and time which I wish to lay before you have sprung from the soil of experimental physics, and therein lies their strength. Their tendency is a radical one. Henceforth space by itself, and time by itself, are doomed to fade away into mere shadows, and only a kind of union of the two will preserve independence.

Bibliography

[1] Stachel, J. (ed.) (1998). *Einstein's miraculous year—five papers that changed the face of physics*. Princeton University Press, Princeton.

[2] Turnbull, H. W. (ed) (1961). *The Correspondence of Isaac Newton*, Vol. 3. Cambridge University Press, Cambridge.

[3] Lange, L. (1885) Über das Beharrungsgesetz. Berichte über die Verhandlungen der königlich sächsischen Gesellschaft der Wissenschaften zu Leipzig; mathematisch-physikalische Classe, **37**, 333–51.

[4] Michelson, A. and Morley, E. (1887). On the relative motion of the Earth and the luminiferous ether. *American Journal of Science* (3rd series), **34**, 333–45.

[5] Shankland, R. (1964). The Michelson–Morley experiment. *American Journal of Physics*, **32**, 16–35.

[6] Brown, H. R. (2001). The origins of length contraction: I. The FitzGerald–Lorentz deformation hypothesis. *American Journal of Physics*, **69**, 1044–54.

[7] Lorentz, H. A. (1885). *Versuch einer Theorie der electrischen und optischen Erscheinungen in bewegten Körpern*. E. J. Brill, Leiden.

[8] Shankland, R. (1963). Conversations with Albert Einstein. *American Journal of Physics*, **31**, 47–57.

[9] Einstein, A. (1905). Zur Elektrodynamik bewegter Körper. *Annalen der Physik (Leipzig)*, **17**, 891–921 and Collected Papers, Vol. 2, Doc. 23. Reprinted in English translation in [1] and [38].

[10] Giulini, D. (2001). Uniqueness of simultaneity. *The British Journal for the Philosophy of Science*, **52**, 651–70.

Bibliography

[11] Minkowski, H. (1909). *Raum und Zeit*. B. G. Teubner, Leipzig and Berlin. Reprinted in English translation in [38].

[12] Einstein, A. (1907). Über die Möglichkeit einer neuen Prüfung des Relativitätsprinzips. *Annalen der Physik*, **23**, 197–8 and Collected Papers, Vol. 2, Doc. 41.

[13] Lampa, A. (1924). Wie erscheint nach der Relativitätstheorie ein bewegter Stab einem ruhenden Beobachter? *Zeitschrift für Physik*, **27**, 138–48.

[14] Terrell, J. (1959). Invisibility of the Lorentz contraction. *Physical Review*, **15**, 1041–5.

[15] Penrose, R. (1959). The apparent shape of a relativistically moving sphere. *Proceedings of the Cambridge Philosophical Society (Mathematical and Physical Sciences)*, **55**, 137–9.

[16] Relativistic flight through Stonehenge: www.itp.uni-hannover.de/~dragon/stonehenge/stone1e.htm.

[17] Other visualizations of SR: www.tempolimit–lichtgeschwindigkeit.de.

[18] Einstein, A. (1905). Ist die Trägheit eines Körpers von seinem Energieinhalt abhängig? *Annalen der Physik (Leipzig)* **18**, 639–41 and Collected Papers, Vol. 2, Doc. 24. Reprinted in English translation in [1].

[19] Bahcall, J.: How the Sun shines. arXiv:astro-ph/0009259.

[20] de Sitter, W. (1913). Ein astronomischer Beweis für die Konstanz der Lichtgeschwindigkeit, and Über die Genauigkeit, innerhalb welcher die Unabhängigkeit der Lichtgeschwindigkeit von der Bewegung der Quelle behauptet werden kann. *Physikalische Zeitschrift* XIV (1913) 429 and 1267.

[21] Brecher, K. (1977). Is the speed of light independent of the velocity of the source? *Physical Review Letters*, **39**, 1051–4.

[22] General information on the galaxy M87: www.seds.org/messier/m/m087.html.

[23] Hubble images of apparent superluminal jet-motion in M87: www.stsci.edu/ftp/science/m87/m87.html.

[24] Mittelstaedt, P. and Nimtz, G. (eds) (1998). Proceedings of the workshop on superluminal(?) velocities. *Annalen der Physik (Leipzig)*, **7**, 585–788.

[25] Kennedy, R. and Thorndike, E. (1932). Experimental establishment of the relativity of time. *Physical Review*, **42**, 400–18.

[26] Ives, H. and Stilwell, G. R. (1938). An experimental study of the rate of a moving atomic clock. *Journal of the Optical Society of America*, **28**, 215–26.

[27] Robertson, H. P. (1949). Postulate versus observation in the special theory of relativity. *Reviews of Modern Physics*, **21**, 378–82.

[28] Mansouri, R. and Sexl, R. (1977). A test theory of special relativity: I. Simultaneity and clock synchronisation. *General Relativity and Gravitation*, **8**, 497–513.

[29] Comprehensive collection of references concerning the experimental basis of special relativity: math.ucr.edu/home/baez/physics/Relativity/SR/experiments.html.

[30] Müller, H. *et al.* (2003). Modern Michelson–Morley experiment using cryogenic optical resonators. *Physical Review Letters*, **91**, 020401.

[31] Wolf, P. *et al.* (2003). Tests of Lorentz invariance using a microwave resonator. *Physical Review Letters*, **90**, 060402.

[32] Michelson–Morley experiment: qom.physik.hu-berlin.de/research_mm.htm.

[33] Kennedy–Thorndike experiment: qom.physik.hu-berlin.de/research_kt.htm.

[34] Saathoff, G. *et al.* (2003). Improved test of time dilation in special relativity. *Physical Review Letters*, **91**, 190403.

[35] Ives–Stillwell experiment: www.mpi-hd.mpg.de/ato/rel/.

[36] Beckman, F. S. and Quarles, D. A. Jr. (1953). On isometries of Euclidean spaces. *Proceedings of the American Mathematical Society*, **4**, 810–15.

[37] Alexandrov, A. D. (1975). Mappings of spaces with families of cones. *Annali di Matematica (Bologna)*, **103**, 229–57.

[38] *The Principle of Relativity* (A collection of original papers by Lorentz, Einstein, Minkowski, and Weyl in English translation). Dover, New York.

Glossary

Aberration: Apparent change of the direction of light due to a relative motion between observer and light source.

Arc minute, arc second, and radian: The 360th part of the full circle is a (angular) degree (denoted by °). The 60th part of a degree is an arc minute (denoted by ′) and the 60th part of that is an arc second (denoted by ″). The $\frac{2\pi}{360}$-fold of an angle, measured in degrees, is its measure in radians.

Dispersion: The phenomenon of the frequency dependence of the index of refraction.

Doppler effect: Modulation of the frequencies of wave phenomena due to (relative) velocities between sources and receivers. Named after the Austrian physicist Christian Johann Doppler (1803–1853).

Ecliptic: The plane in space in which the Earth orbits the Sun.

Ether: Outdated concept of a hypothetical medium, supposed to be the carrier for light and generally all electromagnetic fields. Failed attempts to reveal any motion relative to the ether by means of physical experiments, together with the extension of the principle of relativity beyond the realm of mechanics, triggered Special Relativity. The idea of an ether is inconsistent with Special Relativity, as long as it gives rise to physically preferred systems of reference (e.g. the ether's rest frame).

Event: Physical process that is strongly localized in space and time. In the idealized (and somewhat unphysical) limit of infinite localization an event is identified with a point in space-time.

Field: Association of a (physical) quantity to each point in space-time. This quantity can be a number (plus physical unit), like, e.g. for the temperature field, a vector, like in the case of a force field, or many vectors, like in the case of the electromagnetic (electric plus magnetic) field.

Field theory: A theory in which the fundamental physical quantities are fields. Maxwell's theory of electromagnetism is such a field theory. Typically field theories describe physical systems with an infinite number of degrees of freedom.

Galilei transformations: Mappings of space-time onto itself which implement the principle of relativity in Newtonian mechanics. These mappings in particular preserve the relation of being simultaneous, i.e. events which carry the same label of time will continue to do so after the mapping (the label itself might have changed). This means that Galilean transformations preserve absolute simultaneity.

Inertial system: Denotes originally a spatial reference system relative to which force-free mass points move along straight lines. In modern contexts also used in connection with space-time reference systems, where the time scale must then also be an inertial one. The world lines of force-free mass points are then also straight.

Inertial timescale: Measure of time, relative to which force-free mass points move uniformly, i.e. move equal distances in equal time intervals.

Interference: Phenomenon of local amplification as well as attenuation of the amplitude of superposed waves.

Lorentz invariance: Being invariant under Lorentz transformations. An equation is called Lorentz invariant, if the Lorentz transformations map solutions to solutions.

Lorentz transformations: Mappings of space-time onto itself which implement the principle of relativity in Einstein's adaptation of

Newtonian mechanics, Maxwell's electrodynamics, and all other fundamental theories of interactions, except gravity.

Physical dimension: The unit by which a physical quantity is measured, like metre, second, kilogram, or any combination thereof obtained from multiplication and division.

Principle of relativity: Demands the dynamical laws to be the same in all inertial reference systems. Hence no inertial reference system is dynamically distinguished within the set of all inertial reference systems.

Relativistic Quantum Mechanics: Adaptation of ordinary Quantum Mechanics in order to render it Lorentz invariant. The result is physically and mathematically inequivalent to ordinary (Galilean invariant) Quantum Mechanics.

Relativistic Quantum Field Theory: Lorentz invariant quantum theory of fields, like Quantum Electrodynamics—the quantized version of Maxwell's theory.

Simultaneity: Needs to be defined for spatially separated events, usually through some procedure to synchronize a spatial distribution of clocks. Otherwise there is no spatially extended notion of 'time' with respect to which 'equality' can be asserted. The phrase 'relativity of simultaneity' largely refers to this dependence on a synchronization procedure.

Space-time diagram: Diagrammatic representation of a (physical) process in space and time, usually with reference to special systems, like inertial ones.

Test theory: A general class of theories containing free parameters (or entire functions), which for special values of these parameters reduce to the theory (here SR) to be tested. Test theories are needed in order to make meaningful quantitative statements concerning the experimental status of a theory.

Time: What one reads off clocks.

World line/surface etc: Represents the motion of a point (extensionless object) or a line (one-dimensional extension) etc. in a space-time diagram.

Symbols, units, constants

Throughout we use SI units, based on the metre (m) for length, the second (s) for time, and the kilogram (kg) for mass.

β	velocity parameter	$\beta = v/c$
γ	dilation factor (γ-factor)	$\gamma = \gamma(v) = 1/\sqrt{1 - v^2/c^2}$
km	kilometre (length)	km $= 10^3$ m
AU	astronomical unit (length)	AU=149 587 870 km
ly	light year (length)	ly $= 9.454 \cdot 10^{12}$ km
Å	Ångstrøm (length)	Å $= 10^{-10}$ m
J	Joule (energy)	J $=$ kg $\cdot$ m$^2 \cdot$ s^{-2}
eV	electron volts (energy)	1 eV $= 1.60210 \cdot 10^{-19}$ J
MeV	mega-electron-volts (energy)	10^6 eV
GeV	giga-electron-volts (energy)	10^9 eV
c	speed of light in the vacuum	$c = 299\,792.458$ m $\cdot$ s^{-1}
$h, \hbar$	Planck's constants	$h = 2\pi \cdot \hbar = 6.626 \cdot 10^{-34}$ J $\cdot$ s
α	fine-structure constant	$\alpha = 1/137.036$

Picture Credits

Fig. 2.1 R. Sexl and H. Urbantke. *Relativity, Groups, Particles*. Springer Verlag (Vienna, 2001). There printed as frontispiece.

Fig. 3.16 J.D. Jackson, *Classical Electrodynamics*, second edition, John Wiley & Sons, New York (USA), 1975. There Fig. 13.4 p. 629.

Fig. 4.2 Taken from internet page hyperphysics.phy-astr.gsu.edu/hbase/nucene/nucbin.html

Fig. 4.2 Background picture taken from internet page: www.cerncourier.com/main/article/43/6/14/1/cernbub3_7-03

Fig. 4.4 N. Ashby: *Relativity in the Global Positioning System*. Living Reviews lrr-2003-1. There figure 2. Online available via
relativity.livingreviews.org/Articles/lrr-2003-1

Fig. 5.4 W. Panofsky und M. Phillips, *Classical Electricity and Magnetism*, second edition, Addison-Wesley (Reading Mass., 1962). There Fig. 22.4 p. 414.

Fig. 5.7 From the PhD-thesis of Guido Saathoff, University of Heidelberg 2003. Courtesy of Dr Guido Saathoff.

Index

aberration 22–4, 63–5, 76, 139–41, 158
  and conformal transformations 141–2
Airy, George Biddell 24
Alexandrov's theorem 150
Annalen der Physik 1, 75
antiparticles 94–5
atomic number 88
atomic physics
  applications of special relativity in 86–8

Balmer series 129
Balmer, Johann 129
Beckman and Quarles theorem 146, 150
Bessel, Friedrich Wilhelm 24
big bang 124
binding energy 89–90, 91
black holes 105, 106, 119
Boltzmann, Ludwig 9
boost *see* velocity transformation
Born, Max 34
Bradley, James 23, 24
Brownian motion 1

Cassini, Giovanni Domenico 110
causality relations 60–3
clock synchronization 45
  by clock-transport 42–3, 137–9
  Einstein's 43–4, 45, 137

clocks 13, 42
clocks, satellite
  relative deviation from clocks on Earth 99–101
clock-transport, slow 45, 136, 137–8
COBE
  see cosmic background explorer
conformal transformations
  and aberration 141–2
Coriolis force 19
cosmic background explorer 124
cosmological constant 108–9
cosmological microwave background radiation 123–4, 136
Coulomb's law 8

dark energy
  see cosmological constant
dark-matter problem 109
de Coulomb, Charles Augustin 8
de Sitter, Willem 114
de Sitter's experiment 114–5, 116, 117
deformation hypothesis 7, 39, 125
Dialogue Concerning the Two Chief World Systems 12
Dirac, Paul 95
dispersion 27, 120, 121, 158
distance function
  see metrics
'*Does the Inertia of a Body Depend on its Energy Content?*' 75

Doppler effect 65–7, 76, 158
  longitudinal 127, 129
  transverse 67, 127–8, 129
Doppler spectroscopy 133, 134
γ draconis 24
drag coefficient 26, 60

earth orbital velocity 33–4
Einstein synchrony 43–4, 45–6, 136
Einstein, Albert 1, 2, 34, 38, 39, 81, 82, 127
  Nobel prize 1
  PhD thesis 1
electric fields 8–9, 82, 83–5
  see also electromagnetic fields
electric forces 8
electrodynamics of moving bodies 1–2, 21
electromagnetic fields 9, 20, 82
electromagnetic induction 39–40
electromagnetism 2, 21
electron-positron pairs 95–6
elementary particle physics
  applications of special relativity in 92–8
elementary particles 94, 96
emission theories 114, 115, 117
energy content 75
  effect on inertia of a body 75–9
energy velocity 121
equinoctial hours 13–4
equivalence relation 44
ether 6–7, 11, 20–1, 38, 41, 124, 158
'*Ether and the Earth's Atmosphere, The*' 35
ether system 66, 82, 122–3, 124, 130
ether theory 26, 31, 33, 40, 114, 129–30
ether velocity 33–4
ether vortices 9
ether wind 21, 30–1, 32, 33, 35
ether's rest frame 20, 40, 158
Euclidean distance 50, 51, 146, 150

Euclidean distance function 146, 147
Euclidean geometry 51, 150, 151
Euclidean metric 145
Euclidean motions 146, 150
event 14, 16–7, 44, 56, 62, 144, 158–9

Faraday, Michael 8
field lines 8
field theory 159
fine structure 87
FitzGerald, George Francis 2, 35
FitzGerald-Lorentz deformation hypothesis 35–7
Fizeau, Armand Hippolyte 25, 26
Fizeau's experiment 25–8, 30, 38
  application of addition law of velocities to 60
Flügge, Siegfried 91
force, transformation laws for 144
Fourier component 120
Fourier composition 120
Fourier, Jean Baptiste Joseph 120
Fresnel, Augustin Jean 20, 26
Friedmann equations 108
front velocity 121

galaxy M87 119
Galilean transformations 14–5, 18–9, 49–50, 81, 159
Galilei, Galileo 12, 110
general relativity 2, 19, 99, 100, 105
geodesy 3
Gibbs, Josiah Willard 9
global navigation satellite system 98, 101
global positioning system 3, 98–9, 101
GLONASS
  see global navigation satellite system

GPS
  see global positioning system
gravitational collapse 105–6
gravitational force 7–8
group velocity 120, 121

Heaviside, Oliver 35
hydrogen 86, 87, 91
hydrogen, atomic 129
hyperfine structure 88
high resolution saturation
  spectroscopy 133, 134–5
Hertz, Heinrich 2, 9, 10

inertia
  effect of energy content on 75–9
inertial mass 79
  impact of material stress on
    107–8
inertial system 14–5, 19, 42, 59, 159
inertial timescale 14–5, 159
'Inquiry into a Theory of Electrical and Optical Phenomena in Moving Bodies' 39
interferometer 29, 32, 34, 122, 123, 125, 133
Io 29
  orbital periods 110–1
IS experiment
  see Ives-Stilwell experiment
isotopes 89
Ives, Herbert 127
Ives-Stilwell experiment 37, 67, 127–30, 131
  modern version 133, 134–6

Joos, Georg 35
Jupiter moons 110

Kennedy, Roy 123
Kennedy-Thorndike experiment 37, 122–7, 130, 131, 132

kinetic energy
  as a function of velocity 74–5
  mass as a function of 75–9
KT experiment
  see Kennedy-Thorndike
    experiment

Lamb shift 88
Lampa, Anton 69
Lange, Ludwig 14
law of addition of velocities 15–6, 28, 38–9, 63
  modification of 58–9
  application to Fitzeau's
    experiment 60
law of energy conservation 78, 100
law of inertia 12, 13, 14–20
length contraction 54–8, 93, 130
  reciprocity of 55, 57–8
  and visual appearance 67–9
light propagation 21–2
light, velocity of 7, 27–9
  isotropy 122
  Rømer method of measurement
    28, 110–4
  universality of 3–4, 45, 65
light, wave theory of 6–7
light-quantum hypothesis 1, 76
Lorentz boosts 151
Lorentz contraction
  see length contraction
Lorentz force 40–1
Lorentz group 152–3
Lorentz invariance 81, 83, 95, 159
Lorentz transformations 46–51, 59, 62, 63, 80, 81, 138, 142, 146, 147, 150, 151, 152, 159–60
Lorentz, Hendrik Antoon 2, 35, 39, 81
Lorentz-FitzGerald contraction
  see length contraction
Lorentz-Larmor theory 129–30

magnetic fields 8–9, 10, 82
  see also electromagnetic field
mass 71, 72, 73–4
  as a function of kinetic energy
    75–9
mass defect 89, 91
mass number 88, 89–90
mathematical point 16–8
matter 94, 96
  dualistic concept of 11
Max-Planck-Institute (Heidelberg)
    133
Maxwell, James Clerk 8, 9, 28
Maxwell's equation 35, 38, 40, 114,
    153
  invariance of 80–5
Maxwell's mathematical formalism
    theory 8–9
mechanical principle of relativity
    12–3
Medicean Stars 110
metrics 145, 146
Michelson, Albert Abraham 29
Michelson-Morley experiment
    29–34, 36, 37, 38, 39, 122–3,
    130, 131, 132
Miller, Dayton 34
Minkowski diagrams
  see space-time diagrams
Minkowski, Hermann 60, 145, 153
Minkowskian distance 147–50, 151
Minkowskian geometry 150, 151,
    152–3
Minkowskian squared distance 148
MM experiment
  see Michelson-Morley experiment
molecular vortices 10
momentum 70
  conservation of 70–3, 76, 79
  in special relativity 70, 73
  transformation laws for 142–3
Morley, Edward 29
'moving clocks slow down'
  see time dilation

muons 93
  detection rate on Earth 93–4
μ-mesons
  see muons

navigational systems 98
  applications of special relativity in
    3, 98–102
neutrons 88, 89
Newton, Isaac 5, 7
Newton's constant 17
Newtonian force 70, 144
Newtonian gravity 105, 106
Newtonian mass 73
Newtonian mechanics 5–6, 11–2,
    70, 92
nuclear fission 91–2
nuclear physics
  applications of special relativity in
    88–92

Oppenheimer-Volkov equation 108
orbiting clocks
  see clocks, satellite

partial wave
  see Fourier component
particle accelerators 92
particle theory of light 6
PCT theorem 97–8
Penrose, Roger 69
phase velocity 120, 121
phenomenological theory of gases
    9–10
*Philosophiae Naturalis Principia
    Mathematica*
  see *Principia, The*
photoelectric effect 1
photons 76, 95–6
Planck constant 17, 76
Planck length 17
Planck scales 97

Index

Planck time 17
Poincaré, Jules Henri 2, 81, 152
point masses 5–6
positrons 93, 95, 96
Principia, The 5, 6, 7, 12
principle of relativity 160
  in electromagnetism 20–1, 41
  in mechanics 11–20, 21
protons 88, 89, 95
Proxima Centauri 24, 104, 105
pulsar 117

quantum chromodynamics 91
quantum electrodynamics 88
quantum field theory 80
quantum fields 96–7
quantum gravity 17
quantum mechanics 86

reference system 11, 13, 19–20
  see also inertial systems
relativistic corrections 86–7
relativistic quantum field theory 96–7, 160
relativistic quantum mechanics 80, 160
rest mass 74, 75, 79
rigid body 5–6
Ritz theory 114, 115–6
Ritz, Walter 114
Rømer, Ole 6–7, 110

scaling maps 141, 142
Schrödinger's equation 86, 87, 95
sidereal periods 110
signal transmission 120–1
simultaneity 2, 4, 19, 41–6, 160
solar system 124
  speed of 28–9
space travel
  applications of special relativity in 102–5

space-time diagrams 16, 50–1, 56, 57, 60, 160
space-time distance 144, 146
space-time relations 2, 3–4
special relativity 1–2, 20, 57, 79–80, 130–1
  applications 2–4
    in atomic physics 86–8
    in elementary particle physics 92–8
    in navigational systems 98–102
    in nuclear physics 88–92
    in space travel 102–5
spin-statistics theorem 97
star parallax 24
stereographic projection 140, 141
Stilwell, G. R. 127
Sun
  radiation power of 90–1
superluminal velocities 117–22
synodic period 111

time dilation 51–4, 93, 100–1, 126–7, 136, 138–9
  reciprocity of 53–4
temporal hours 13
Terrell, James 69
test theories 131, 133, 160
Tevatron ring 92
Thorndike, Edward 123
tidal friction 19
timescale 14, 19
  see also inertial timescale
transformation laws
  for force 144
  for momentum 142–3
transformations 15, 150, 151

Universe
  accelerated expansion of the 97, 108–9
uranium 91–2

vacuum 96
  fluctuation energy 96–7
velocity transformation 14
Virgo cluster 119
Voigt, Woldemar 2, 81

world line 16, 17, 46, 47, 53, 56, 160
world surface, rod's 55–6

Young, Thomas 6

Zeeman, Pieter 26–7

EUGENE PICKENS